T0105628

Is Your Fork In Tune?

The Search for Resonance

Hayley Weatherburn

ISBN: 978-1-4525-5028-2 (sc)
ISBN: 978-1-4525-5027-5 (e)
ISBN: 978-1-4525-5029-9 (hc)

Balboa Press books may be ordered through booksellers or by contacting:

Balboa Press
A Division of Hay House
1663 Liberty Drive
Bloomington, IN 47403
www.balboapress.com
1-(877) 407-4847

Printed in the United States of America

Library of Congress Control Number: 2012907496

Balboa Press rev. date: 05/03/12

To Mum and Dad, my two polar opposites:
thank you for showing me everything and for
allowing me to make up my own mind.

Thought is real. Physical is the illusion
—What Dreams May Come

Contents

List of Illustrations

Preface

If someone had told me five years ago that I would finish a book of more than fifty thousand words about quantum physics and the answers to the meaning of life, I would never have believed it. I started writing all of this down with the loose notion that it might turn into a book, but more with the idea of working out the thoughts in my head. I was at a loss when it came to what I believed in. I found it difficult to figure out what I wanted to do with myself when I didn't know what I stood for and why I was here. So in an attempt to find out what I believed in, the first drafts of this book emerged.

It was an on-and-off process. Sometimes I just researched or took courses to absorb knowledge. At other times, I simply went through periods of subconscious analysis when, without focusing on any concepts, my mind spit out some gems seemingly out of nowhere. After collating my many notes on numerous pages, I began writing my first draft. After a couple of years of trying to put it all together, I had had enough. I felt at a loss for ideas, and I was getting nowhere. So I decided to ignore the project for a year.

I didn't last a year. After eight months, the burning desire to work everything out was too strong, and I decided to reopen my first draft and dive into it again. Balancing my life between work and the book, it took me another two years to finish. I am glad I persevered, because I now have a clear understanding of what I believe and a clear direction as to where I am headed.

I have many people to thank for their help along the way, all of whom I acknowledge on the next page. However, I just want to put out a general thank you to everyone who crossed my path in these last four years, since you all had something to contribute. Whether you made a random comment that triggered an answer for me or inspired me through a direct conversation about metaphysics, everything contributed to this final edit—so thank you.

Acknowledgments

I am truly grateful to all my family and friends who have supported me along the way. This book has taken four years to come to fruition, and there was someone encouraging me every step of the way.

I want to first thank my mum, who introduced me to books that stretched my mind and taught me to think a little differently from others. She is an amazing writer and my inspiration to this day. My dad, who is an excellent devil's advocate, made me think about the things I didn't want to, and as crazy as my ideas may have sounded, he still supported me. Thanks also to Tristan, my brother, for all our mind and energy conversations that helped me come to my ideas. To my other brother, Daniel, thank you for completely and utterly believing in me. In fact, as a young boy, you absorbed these beliefs and truly live the life of resonating with your reality! To Erin, thank you for volunteering your talents for the drawings; you understood my concepts and what I wanted and really brought them to life for me.

To the best bosses in the world, Chris and Melanie Hassall of Hassall Auctions—thank you for your support and encouragement of my crazy book ideas. Your support is truly appreciated, and I couldn't have done this without your help and understanding.

And, of course, thank you to all my friends in Brisbane, Byron Bay, Newcastle, Canberra, Sydney, and around the world who let me talk about these ideas and who said, "That book sounds interesting—I can't wait to read it!" Whether you meant it or not, it really helped me to finish it. I do hope you all enjoy it.

Before I embarked on my journey of learning about how a book comes together, I never realized just how much work goes into the editing, formatting, and finalizing of it all. I couldn't have done this without a few people I must mention! I want to thank them for their hard work, time, and dedication in helping this book become the best it could be!

Doreen Zeitvogel, thank you for the hours of professional work you put into editing this book—it will be forever appreciated! You truly were an angel sent from the universe to help make it become a real book!

Nicole Hilton, thank you for editing the index—again, another part of a book I think we all take for granted. Without the index, my research for this book would have taken a lot longer. Thanks for putting my index together and adding it to the book's final edit.

Belinda Millane, your brilliant skills as a graphic designer transferred over to book covers with ease! You brought to life the image that had been stuck in my head and made it a reality. Although the final cover design was not exactly yours, it would never have got to this stage without your effort and hard work. I am forever grateful Thank you!

Ben J. Woods, thank you for your generous support in assisting me in the promotion of this book.

Chapter 1: So Many Questions

INTRODUCTION

Life is like a tuning fork, and the answer to everything is energy. The end.

I hope you enjoyed the book. I really don't need to go on. I've given you all the revolutionary answers you need to know to change your life and bedazzle your friends and family. I imagine you feel inspired, revived, and satisfied with knowing the answers to all those unexplained phenomena you've always wondered about, including the meaning of your life. You are seeing the world through different eyes.

For those of you who prefer a bit more information to back up this amazing find, I guess you could choose to read on. Perhaps you'd prefer a few real-life stories to add interest, or maybe a bit of fascinating science to add substance, or even some interesting research thrown in for good measure.

So choose to read or not to read: it's up to you. However, I recommend that you read further—and don't just believe the first thing you read.

Let me start from the beginning. Why did I even begin to search for answers to questions that, until recently, have vexed me for most of my life?

As a curious young thing, I was always asking why. As a kid, much to my parents' frustration, I would follow any comment with an inquisitive "Why?" over and over again. A simple conversation could go like this:

Mum: "Hayley, pop in the car. We are going to the shops."
Me: "Why?"
Mum: "We need some food."
Me: "Why?"
Mum: "Because we need to eat."
Me: "Why?"
Mum: "If you don't eat, you get sick."
Me: "Why?"
Mum: "That is just how the body works."
Me: "Why?"

Mum: "Just because! Now please stop asking questions and get in the car."

I am surprised my parents didn't literally strangle me during one of these conversations. A simple statement about putting on my socks could elicit all kinds of questions about the meaning of the universe!

Throughout my life, this curiosity and thirst for answers has never stopped.

At school, we were taught a vast array of subjects, some of which were contradictory. Take science and religion, for example. My favorite subjects were science and math. Why? Because there were only right or wrong answers. It seemed simple enough to learn these facts—until I went to religion class and was told different ones.

This completely confused me. How could one subject tell us one thing—that the universe began with a Big Bang—and then another tell us that God created the universe and everything in it? How could teachers and parents (at that time, the authority figures in my life) teach us two completely contradictory answers? How could they both be right? And if they couldn't, then which one should I believe? Science, which had the facts, or religion, which asked you to rely on faith? I certainly hated being wrong, so what could I do? So throughout my time in school, I decided to accept both tentatively and work out later which was right. This conflict remained in my subconscious, keeping me intrigued by these areas of questioning until I eventually decided to become a little more proactive about it all.

In the meantime, there were other issues that caught my attention.

At school, I don't think I ever really had a sick day. I often heard my mum say to other parents, "Hayley never gets sick." And I didn't. Then there were other kids who *always* got sick. I thought I was just lucky. Growing up, I learned about the "placebo effect". This is when a doctor gives you a tablet or other type of medicine that actually has no drug in it at all, but the doctor tells you it fixes your ailment. Surprisingly, it still often works and can heal without any medicine involved. This concept of miracle healing intrigued me. I heard of people becoming better overnight—which baffled me—or worse, becoming intensely ill overnight. Could the mind heal or destroy matter? This phenomenon lodged in my subconscious next to my previous unresolved conflict, both awaiting some kind of answer. Only instead of finding solutions, I continued to add to the ever-growing list of questions I had filed away in my mind.

From the time I was six years old, my parents took me traveling. By age twelve, I had been to Los Angeles, Mexico, Singapore, Hawaii, Malaysia, and Hong Kong, and I had even lived on Christmas Island in the Indian Ocean for two years. In the process, I was introduced to many different cultures with all their interesting beliefs and superstitions. Christmas Island was mostly populated by Chinese and Malaysians. We celebrated the Chinese New Year, and I remember getting red envelopes with money in them. Red meant good luck, and the money was supposed to bring you prosperity—at least, that's what I remember understanding as a child. Down in the Malay *kampong*, or village, I also remember hearing Muslim prayers continually being announced over a loudspeaker and watching Muslim worshippers praying on mats that faced east.

These different influences opened my eyes to new cultures, beliefs, and superstitions, adding to my questions about religion. Not only was I now asking which was right—science or religion—but also which religion was the right one. They seemed to have similarities in that they all had a God or gods, but they also had a lot of differences in their beliefs and superstitions. Later, I discovered that even some of the sciences conflicted with each other. The whole ordeal bothered me, but at this stage it wasn't yet enough for me to do something about it. I could still live life well enough without resolving these issues.

At the age of twenty-one, I was well on my way to achieving what was expected or appropriate in my society. I had graduated from university, found a full-time job, bought a car, and was running an internet business on the side in an effort to build my income for an early retirement. Although by all accounts I should have felt satisfied, instead I felt hollow, empty, and directionless. Based on a brilliant suggestion from my mother, I made the decision to travel overseas for a year, and within three months, I had packed my bags and left. What was to be one year turned into five years of traveling throughout Europe, Africa, Asia, and North America. On the way, I had many amazing experiences and met a number of unique individuals who shared their experiences with me. During this time, I accumulated more and more unresolved questions to store in my to-be-answered-later file.

One great paradigm-shifting experience happened to me in France. While staying in an old chateau, a friend and I both experienced a haunting. I'll speak of this in detail later in the chapter; for now, I just wanted to bring it up to explain the conflict I was feeling at the time. There I was, with more of a mind for science than for what I felt was fiction, being haunted. It didn't make sense to me, so I added it to my ever-growing file and continued on my journey. Talking about this experience to others only made me aware of more unexplained phenomena: stories of animals seeing ghosts, of people randomly thinking of their friends back home and then bumping into them in a different country not long afterward, of psychics who could talk to people who had passed away or even predict the future. I crammed all this into my now overflowing file.

At the age of twenty-six I decided to come home. I was feeling the urge to build a life that would allow me to travel anywhere, anytime without worrying about money. I also wanted eventually to have my own family and didn't think I would find someone to do this with me while I was traveling and leading a very unpredictable life. This decision finally led me to do something about all the unanswered questions that had plagued me for so long.

During my time in sales, I read many books about the power of the mind. These were books that inspired you to do better, that tapped into the power of the human brain, and the ideas in these books fueled my interest in the potential that belongs to all of us. Considering that it had been two years since I had come back home and that I still hadn't found a job that really ignited my passion, I decided to start looking into all those unanswered questions. I was curious. I earned a life coaching certificate and started a part-time business. This only enhanced my interest in human potential, and it was during this time of part-time work and part-time coaching that I found the time to start researching, reading, taking courses and seminars, and watching documentaries. Ideas and even some answers started to trickle in. Each idea developed over time and grew like a flower blossoming in a field of potential answers. In 2007, I started to write these ideas down, essentially to work out what I believed in. What, I wondered, were my answers to these questions?

Over four years, this journey that started off as a quest to answer a nagging file of questions turned into something that, to me, is pretty exciting—an answer or theory that responds to a number of intriguing questions of varying depth and scope. How is it that people often randomly think of someone just before that person calls? How can religion and science both be right? How can a superstition or belief be meaningful to one person while the opposite belief is true in another person's experience? In other words, how can evidence exist for opposite beliefs?

I now feel I can answer these and other questions. I am excited about what I have discovered and am even more excited to share it with so many people. I honestly believe that there is something in this theory of mine. I feel it right down to the core of my being. It could very well contain some ideas that are wrong, but I feel that the reality that it's uncovered for me fits who I am. It is my box, and though it may not fit you, if it helps you to discover your own answers along the way, then I will be happy.

For the rest of this chapter, I'll be elaborating on the different topics that pertain to the questions I have been grappling with, and I'll give some examples so that you can understand where I am coming from. The second chapter will delve into the answer, the theory that resulted from researching different areas of science. Although I deal with the realm of quantum physics, don't be concerned if you're not a scientist, as I've broken it down into simple explanations to help clarify the complete theory. Don't worry if you don't understand each part fully at first, because in the third chapter, I'll illustrate and explain the theory through real-life examples. The fourth chapter will help you to use this concept to your advantage in daily life. Finally, in the last chapter, I will take the theory even further into deeper levels.

Let's start by elaborating on those questions I had rolling around in my subconscious. I have separated them into subheadings to explain my experiences with each one. I won't yet go into the research in this chapter; rather, I'll concentrate on the experiences, stories, and influences that caused me to start looking into these areas in the first place.

Let's start with something simple and interesting, something that I believe most people have experienced at some time in their lives. Let's explore unexplained connections.

UNEXPLAINED CONNECTIONS

Random Coincidences with People

Have you ever randomly thought of someone you haven't seen in years, only to have them call you five minutes later? This kind of thing happens to me quite a bit. At the Caxton Street Seafood and Wine Festival in Brisbane, Queensland, in 2010, I was walking down the street among a crowd of thousands. I had this strange thought of an old colleague I had worked with in Newcastle, NSW (a town over eight hours' drive away), whom I hadn't seen in over a year. Literally ten minutes later, I bumped into her among the thousands of people there! You don't normally find friends you are supposed to catch up with in those crowds, let alone someone from another town you just randomly thought of. It was uncanny. I have had friends tell me similar stories, so I know I'm not the only one who has experienced this kind of thing.

Another interesting phenomenon of a similar nature is the connection twins have to each other. I met one girl, Cynthia,* who told me a story about how she was hanging out with one of her friends who was a twin. The twin was complaining about her left ankle hurting, that there were really sharp stabs of pain all over her foot and ankle. Cynthia later found out that the other twin had smashed her foot through glass and had glass shards in her foot and ankle. Although this story may have been hearsay, I have heard enough of these stories to pique my interest in the subject, and as a result, I decided to look into it. There must be something, I thought, that could explain this.

These connections don't just happen to twins—they can happen between parents and their children as well. Have you ever heard of a mother suddenly feeling the need to call her child because she knows something is wrong? There are probably more stories of mothers worrying when nothing is wrong, but again, there seem to be enough stories of the other variety to validate looking into the subject. In fact, I have actually experienced this myself.

As I mentioned before, when I was about eight, I lived on Christmas Island in a really awesome, massive house. My room was at the front of the

---

* Not the person's real name

house a little apart from any other area. You wouldn't really walk past it ordinarily. You would only come my way if you actually decided to come and see me. Well, one day I was being my usual curious self and had my bedside lamp in my hands. It was still plugged in, and I decided I wanted to see how it was put together, so I unscrewed the lightbulb. I saw the two prongs at the bottom and thought I would touch them. Just as my finger neared the electrified prongs, my dad came into the room. In an instant, he saw what I was about to do and screamed at me to stop. He told me what would have happened, and I was in a bit of a state of shock. I asked Dad why he had even come into my room to see me. He said he didn't know but that he had just felt a need to come and see me. Coincidence or not, this incident fueled my curiosity about these types of events.

Long-term partners also experience this. I have friends who are couples and who often tell me of instances when they have texted each other at the same time asking the exact same question. Are there enough experiences of this kind to convince us that this is at least worth looking into? I think so.

Energetic Connections

A different type of connection involves the different relationships we have with other people. These are the energetic connections we all feel. Have you ever been with a friend and just felt drained while you were with the person? Subconsciously, you even try to avoid such people, because you know you will be tired afterward. What about the other end of the scale? Have you ever been with someone who made you feel energized, pumped up, and just really motivated or positive? The time you spend with that person seems to fly, and you feel so connected you even forget about the outside world. These types of energetic connections made me wonder what might be going on for this to happen. I mean, how can you talk to someone for ten minutes and be tired afterward when talking to someone else for thirty minutes leaves you with more energy than you had before? The sleuth in me wondered if there was more to this than met the eye.

Then there is the love connection. What makes you love one person so much that you would risk your life for them? Sometimes this connection is completely irrational: the person you love isn't the normal stereotype of who you're attracted to and doesn't have similar likes or dislikes, yet this love connection seems to hypnotize you. This type of connection can be for an animal or an inanimate object, like a photo or a car. It could even be for a sports team. I have seen the toughest, most unemotional men cry out of happiness when their team won the championship game, yet when these men broke up with their longtime girlfriends, they shed no tears.

This love connection can make people do silly things—it can even make people change who they are. I wondered what was going on to make this happen. I have been on both ends of the scale in different relationships. In one relationship, I liked the guy, but my connection to him was not as strong as his was to me. He seemed to change. He wasn't the individual I was originally attracted to: sporty, funny, with his own agenda and personality. He eventually just wanted to do whatever I was doing. He never really had any ideas of his own, and he would do whatever made me happy. I couldn't even provoke him into arguing so he could demonstrate some of his individuality. It was frustrating.

Then I experienced it for myself. I quickly fell for a guy who became my boyfriend. I felt totally connected and would get butterflies when he spoke to me or held my hand. The connection we had hypnotized me, and I was terrified of losing it, so I did whatever he wanted me to do even though one side of my mind was telling me, "Don't do that; that's not who you are!" The other side that was hypnotized by this connection just floated along, following him, sort of like those cartoons of people in love hovering above the ground.

Then, of course, there is the ultimate connection when both partners feel the same level of love. It's a feeling of euphoria—an ecstasy that is rarely matched in any other way. Yes, there is a degree of biology and chemistry involved, but sometimes it feels like more than just that—like when my couple friends think about the same thing and text it to each other at the same time. Other people might finish each others' sentences or know exactly what the other person is thinking. Is this just a matter of knowing someone's behavior patterns really well? Or is there actually a deeper level of connection taking place that allows you to practically read the other person's mind?

Passionate Interests

A different sort of connection is the connection to an activity. What I mean here is that sense of connection you feel when you're doing something you enjoy or love. Examples of this might be top athletes playing a sport they are passionate about, dancers doing their favorite dances, opera singers in the midst of song, builders creating masterpieces, auctioneers skipping through their lyrics during a sale, farmers growing a strong crop or herd, stockbrokers making their trades, mothers watching their children achieve—I could go on and on, but I think you understand what I mean by having that connection to an activity. If I swapped the farmer with the singer, do you think the singer would feel the same connection to the crop or the farmer to the song? Yes, they might also have a small connection to these things (we are human and can be interested in more than one thing); however, we are usually more connected to one thing than to another. It's this type of connection I am talking about—the kind that creates a feeling of elation, love, high energy, and satisfaction.

Why did these connections pique my interest? I found it fascinating to watch someone be so passionate about one thing. What was it that drove people to do crazy things to achieve their goals of connecting with these activities? Why did athletes push themselves to be the best at their sports? Why do people on shows like *So You Think You Can Dance?* cry out so much to be a part of dancing? The only thing that matters to them in life is to be able to dance. I was intrigued by this connection, or even obsession, that could create that elated feeling they were chasing. But what about those people that felt they had no connection or weren't sure what their connection was? How might they find this passion? These questions and more made me look into these unexplained connections.

Questions about unexplained connections I will answer throughout this book include the following:

1. How is it possible that twins, family, or even random people can be connected?
2. Why do some people experience this synchronicity more than others?
3. How can connecting with one person make you feel drained while connecting with another energizes you?
4. What is going on, other than biology and chemistry, when there is a love connection?
5. Why do we connect with some activities and not with others?

STRANGE ANIMAL BEHAVIOR

Connections

I had a beautiful, blue-eyed Husky named Skye. She was very smart. Even when my family spelled it out or said it quietly to each other, she always knew when she was about to be walked. I've heard amazing stories of friends' pets knowing whether they were going to the vet. Most of the time they loved going for drives; however, when it came to the drive to the vet, they would refuse to get in the car. I used to speculate about how these animals could sense this. Could they read our minds?

In his book *Dogs That Know When Their Owners Are Coming Home,* Dr. Rupert Sheldrake quotes a story by Veronica Rowe about a cat that responded to the telephone whenever a particular person called:

> Seven years after she acquired Carlo, my daughter went to teacher training college and rang us infrequently. However, when the phone did ring and it was Marian and not our son, who was away at Kingston Polytechnic, Carlo would bound up the stairs (the phone was on the half landing) before I had picked up the receiver! There was no way that this cat could have known my daughter was to ring us—it was a standing joke when he bounded up the stairs that Marian was on the other end of the phone. He never did this at any other time and was not allowed upstairs anyway.[1]

Was this cat reading the mind of the caller, or could it just predict who was going to call? How did it know? Could it simply be that the owners had selective memories? I suspected the latter but couldn't discount the other possibilities, especially when I read of other people's experiences.

Have you ever heard stories about animals just knowing what was going on around them through some kind of intuition, such as knowing exactly when you were on your way home? Another interesting story I read in Sheldrake's book was experienced by Elizabeth Bryan: "My whole working life has been as a cabin crew member working out of Gatwick Airport. For ten years my dog Rusty would jump around and bark at the same time I landed and then sit quietly watching the front door until I got home. The astonishing thing is there is no routine to my coming and goings—I could be gone one day or fourteen and no regular time of landing, yet he knew without fail."[2]

You may have experienced or heard from friends of animals finding their owners after they have moved or going long distances to find their owners in different suburbs or even cities. One story that totally blows my mind is of a dog finding his owner's grave:

> My father-in-law had a small farm, and on it he kept a watchdog, Sultan. One day my father-in-law became ill and was taken to hospital by ambulance. A few days later he died and then he was buried in the local graveyard, five kilometers from the farm. Several weeks after the burial the dog was not seen for days. This seemed strange to us, as Sultan never used to stray. But we did not make much of it until one Sunday a former employee came along who lived near the graveyard. She told us: "Imagine, when I went across the graveyard the other day, Sultan lay at your family grave." I cannot fathom how he could have found the way, all these five kilometers. There were no footprints of his former master that he could follow. And he had never been taken to the graveyard, not even to the fields, since he had to keep watch at the house. How is it possible that he found his master's grave?[3]

Stories like that encouraged me to keep looking further into these mind-boggling questions! And these animal stories weren't just limited to connections: they delved into medical mysteries as well.

Medical Masters

In 2008, a documentary I was watching on dolphins talked about the Harmony Program in Panama City, Florida, which was specially designed for disabled children so that they could swim and play with dolphins. Each child had a different type of disability and could only be played with at certain levels. What was amazing was that these animals always seemed to know the children's limitations as to how much they could play. It's as though the dolphins could sense exactly what was different about each child and play accordingly.

Other medical documentaries I've seen report on certain dogs that are able to sniff out cancers, while others are able to predict epileptic fits. Sheldrake cites an example of this: "Steven Beasant of Grimsby, Lincolnshire, is regularly warned of impending fits by his dog Jip, a mongrel. Normally

Jip follows him around and stays very close prior to an attack, and when Steven is sitting down, the dog jumps up on him. But Steven says that Jip sometimes 'comes bounding through from the kitchen and then he will pin me to the chair.' So whatever signals Jip is reacting to can be felt in a different room."[4]

The fact that the dolphins and dogs could read the medical ailments of these individuals added more questions to my ever-growing list of curiosities. Could this have anything to do with the connections I had previously been wondering about between humans, or was this different?

Psychic Phenomena

Probably one of my favorite interesting stories about pets behaving weirdly involves my friends' dog, Ruffy, a small Jack Russell Terrier. My friends took Ruffy with them everywhere, including their holiday farmhouse in Hill End, NSW, Australia. Ruffy hated going into that place. He would bark or shy away. There was something about it that he couldn't stand. It was a very old farmhouse and probably had a lot of history. I believe that Ruffy could sense ghosts. You may have seen movies like *Ghost,* in which the cat can sense Patrick Swayze's character. Yes, it's just a movie, but the idea stems from people's experiences with animals sensing parapsychic energies.

The idea of animals having this psychic sense doesn't just apply to ghosts but to predicting future events as well. During the horrific tsunami of 2005 in Indonesia, there were stories of elephants running to the hills up to an hour before the tsunami hit. Sheldrake also collates many different reports of animals predicting silent bombings and plane attacks during the Second World War.[5]

From all these different experiences, one can deduce that something is going on at a deeper level that we don't yet quite understand. It was the combination of all this information that made me ponder the possibilities.

Questions about strange animal behavior I will answer throughout this book include the following:

1. How do pets know the intentions of their human owners?
2. How do animals sense the medical ailments of humans?
3. How do animals sense paranormal phenomena, such as ghosts?
4. How is it possible that animals can predict earthquakes, tsunamis, and even bombings?

PARANORMAL ACTIVITY

Ghosts and Guardian Angels

I have been haunted, and to this day I wouldn't believe it if it weren't for the way it happened. I was staying in a chateau in France while on a tour in Europe. It was our second night there, and I was in a room with three other girls. We chatted until the late hours and eventually all fell asleep.

Just as I was dozing off, I felt an immense pressure push down on my chest as though I was being held down. I felt like I was screaming loudly and trying to move, but I knew somehow that no one could hear me. It was like being paralyzed in some sort of vacuum. I thought I heard a deep voice telling me to get out. After what felt like ten minutes but was more likely only about one, the pressure left just as quickly as it had come. My heart was pounding. I didn't really know what had happened, and after calming down a bit, I started to think that maybe I was just deliriously tired and had dreamed it.

Still a little freaked out, I jumped into the top bunk with my friend Sharyn and asked if I could lie up there with her, as I was a little scared. She was barely awake but said it would be fine. After about twenty-five minutes, I had calmed down and was starting to fall asleep again. Suddenly, Sharyn grabbed my arm and said to me, "Did you just hear me screaming? Something was holding me down and telling me to get out!" Well, didn't I just almost pee in my pants! Panicking, I told her that the same thing had just happened to me down on the bottom bunk. That's why I had come up there! We turned the lights on and remained freaked out for the rest of the night!

To this day, I am convinced it was some sort of spirit, but I really don't know how it was possible. The experience must have been real since Sharyn and I experienced it identically without any sort of communication between us. It made me really start questioning whether and how ghosts could exist. I had had friends tell me before that they had seen weird happenings in the chateau—things like taps turning on randomly. A year before, I had worked at the National Film and Sound Archive in Canberra, ACT. While there, I had been told of one particular room that had a poltergeist that moved objects around and that this had been witnessed by a few different people. Several other staff members had also reported seeing a man with a black bowler hat and women wearing old-fashioned clothing. The museum had been used as a ballroom in the 1920s, so this all seemed likely.

Until my haunting, I thought there must be some explanation for these things: people delirious or on drugs, the wind moving things, and so on. But after my experience, which I could not explain, I really wanted to know why and how. Certainly I needed to know more.

Another area of the paranormal has to do with guardian angels. There is a story I heard from my friend Amie when I was about fourteen. She said that once when she was crossing the road, just as she was stepping out off the curb someone pulled her back and a car went screaming past. Her heart racing, she turned around to thank the person, but there was no one there. She believed it was her guardian angel. Hearing this made me question how such a thing could be possible. I had heard similar stories, and they all just fueled my curiosity even further.

Psychic Abilities

In 2008, I went to a psychic. I was quite skeptical but went to this one because, of course, I was curious. I had heard from three separate people who had been to see her and had her correctly predict things that she shouldn't have been able to know, so I decided to go along. She didn't ask me any questions; she just talked for thirty minutes, and I wrote down what she told me.

She told me that angels or spirits were telling her things. There were a few things that even I didn't know but that turned out to be true, which gave me shivers down my spine. She told me about someone being adopted into my family two generations back. It was my great-uncle Frederick, which I found out later when I asked my dad if it was true. She also asked me if I wanted to speak to anyone who had passed. Only knowing one person at that time who had passed, I asked about Little Grandma (who was my great-grandmother, but we called her "Little Grandma" because she wasn't very tall). She said she was there with Margaret. That was actually her mother's name, which I didn't know until Dad told me afterward. There were quite a few other things that she knew, things that you couldn't just pluck from the sky. I know there are people out there who don't believe these things, even if they can't give you a good reason. To them it's just hocus pocus. To my mind, though, there are too many coincidences for there not to be something going on. That is why I kept reading and researching.

Some of this research led me to learn more and more about auras and chakras, which have to do with the energy of the human body. Again, at first it really did sound like hippie talk—about the colors of the body, how chakras can be out of balance, and how your auras can reveal your mood like a mood ring. Now, I don't know about you, but I was very skeptical about it all, and to be perfectly honest, I just put this ability to see energy around people's bodies down to hippies taking hallucinatory drugs. After diving into my research, though, I started reading about scientists proving the existence of energy fields around the body as well as of photographic equipment capable of taking photos of the aura. So, trying to keep an open mind, I added this to my never-ending list of questions to be answered.

Near-Death and Out-of-Body Experiences

In 2010, I was deep into my journey of discovering answers, researching books, and going to different courses. I thought I was beginning to understand it all, and then I came across near-death experiences (NDEs) and out-of-body experiences (OBEs), or astral travel.

There are many documented cases of people who have reported a sensation of standing or floating above their bodies while on the operating table. There are reports of seeing loved ones that have passed or of seeing a light. People who have had OBEs they talk about floating above their bodies near the ceiling, flying through the sky, and even walking through walls. This all seemed a bit of a stretch in terms of reality, sounding more like people's imaginations or dreams. I had certainly had several dreams about flying, but was I outside of my body at those times? I didn't know, but I decided I would like to find out.

My scientific mind couldn't conceive of how this kind of thing could happen, but my curiosity got the better of me, and I started to look into it more. I watched an interview from the 1970s with Robert Monroe of the Monroe Institute. Here was a fifty-year-old talking about his out-of-body experiences.[6] When he first started experiencing OBEs, Monroe thought he was going mad and got a number of doctors to look at him. He thought it was his imagination, but eventually he started to observe and take notes. Originally a businessman who didn't believe in this kind of thing, he stopped everything else and dedicated the rest of his life to researching OBEs, founding the Monroe Institute in 1972 to scientifically study the phenomenon.

Another scientific body that researches this type of thing is the International Academy of Consciousness (IAC), which is where I decided to do an in-depth course called the Consciousness Development Program. I will talk more about my findings later in this book, as you will need the answer/theory before what I say can even make sense. Essentially, though, the idea of people being outside their bodies confused me. The possibility seemed highly unlikely, and it seemed more likely to me that it was simply the imagination of individuals under stressful conditions. Nonetheless, it was another unanswered question that added to my mounting pile of questions, which was beginning to compete with Everest!

Questions about paranormal activity I will answer throughout this book include the following:

1. Could ghosts be real, and if so, how are they possible?
2. What are angels, and how is it possible that they even exist?
3. How are psychic abilities possible? How can the future be read if it hasn't happened yet?
4. What is all this talk about auras, chakras, and the like?
5. Are OBEs and NDEs possible, and if they are, why doesn't everyone experience them?

MIND OVER MATTER

Beliefs

I've always been a big believer in the power of the mind. I think it started when I first heard a story about the aboriginal culture in elementary school. One belief of their culture is called "the pointing of the bones." From memory, if you broke aboriginal law, the elders would get together and discuss your fate. Under the worst circumstances, the elders would decide to point the bones at you—which meant death. Essentially, because of this belief, if this happened to you, you would die soon after the bones were pointed at you—all because of your belief!

This fascinated me, because my own education and upbringing had taught me that such things couldn't happen—that bones were matter and couldn't cause harm to anyone, unless of course they were sharpened and used as weapons. It made me realize that the mind was mysteriously powerful! I wondered what was happening scientifically for a belief to be strong enough to cause someone's death. If our minds were that powerful, what were the other possibilities?

I heard a similar story overseas about a man named Paul (not his real name) who worked for a freezer van company. One Friday, he stayed late at work and walked into the back of a freezer truck to put something in it. At that moment, the back door to the freezer van closed accidentally, and he was locked in from the outside with no way of getting out. After making noises to get someone's attention, he realized that everyone had gone home for the weekend and that he would be stuck in the freezing cold van until Monday morning. After going through shock, he came to accept that he would most likely die in the back of this truck. So to try and make something positive out of a negative, he decided that for the sake of scientific research, he would write down his experience of what it was like to die of hypothermia. At every chance, he wrote the time and what was happening to his body. Eventually, he passed away from hypothermia. On Monday, when they finally found his body and his notes, they were stunned. As with all suspicious deaths, Paul's body was autopsied, and it was determined that he had indeed died of hypothermia. What made it suspicious was that the freezer in the truck had been turned off and it hadn't even been cold. Basically, Paul had made up his mind that he was going to die of hypothermia, so his body did exactly that. Baffling, isn't it?

Hallucinations or Reality—What Is Real?

Less dramatic than the notion of dying but potentially just as debilitating is seeing friends or family so caught up in their own realities that no matter how many times you show them that the freezer isn't on—to use Paul's story as a metaphor—they can't see it. An example of this are people who are convinced that they are fat when anyone can see that they border on being anorexic. The reality that they see is completely different from the reality that everyone else sees. Their perception is like a hallucination; their mind is playing tricks on them. When they look in the mirror, their mind convinces them that the freezer is on, and that is all that they can see or feel. To the rest of us, they look like skeletons—the freezer is obviously off. It doesn't matter how much we tell them otherwise. We can talk to them until we are blue in the face, telling them that they need to eat—they just can't see it to believe it.

There are also times when people trap themselves within a reality created from limiting beliefs. An instance of this might be those people who work extremely hard at being successful but won't accept assistance from anyone because they hold the belief that if anyone helps them, they have failed. What this limiting belief does is create a reality in which it is very difficult to succeed. If they shifted their limiting belief and created a new reality, their success would come a lot more quickly. This idea of individual realities—that each person has his or her own reality—made me wonder what is actually real. Is there one reality that everyone can see, or does each of us see differently?

I mean, how do we really know whether what we see is the same as what other people see? How do we know what experiences they are having? For example, when we were children, we were taught the color green. Every time an adult would point to a certain color and say "green," we learned to associate that color with that word. But how do I know that what I learned to call green is actually the exact same color you see with your eyes? What you see might be my blue, but you call it green. This idea opens a can of worms when it comes to the question of meaning making—the question of what reality is.

Autism

When I was around twelve years old, I remember being on a Sunday drive and asking my parents what autism was. They described it to me as a deficiency that some people have in using the mind's filter. They explained that at all times we have over two million bits of information around us that are being felt, seen, heard, and perceived in other ways through our five senses. The filter that we have in our brains actually cuts this down to around two thousand bits of information felt both consciously and unconsciously.

Autistic people do not have such a well-developed filter, so they are constantly dealing with an unfathomable amount of information. Take, for example, the movie *Rain Man*, which is based on the true story of an autistic fellow named Kim Peek. The autistic character in this film could count the exact number of matchsticks that fell to the ground in a matter of seconds.

There are people out there who can do so much more with their brains—almost like computers! If the statistics are right that we currently use only 10 percent of our brains, what are the possibilities for us in the future? The whole concept of our brains actually taking in so much more than we consciously realize has fascinated me for a long time. I wondered what would be possible if we could really learn to utilize our mind's filters, allowing a lot more to come in. This idea raised more questions for me than I can list here right now. I decided to look into it, and in the coming chapters, I will tell you what I found.

Medical Marvels

Do you know or have you heard stories of people who were told they were going to die within a few weeks or months but then turned around and lived for fifty more years? Then there are other stories of people who have had cancer and have healed themselves without medication. One of the most famous of these is the story of Louise Hay, who healed herself from cancer through mental change.[7]

Another story along these lines that springs to mind is that of Anna Meares, one of Australia's Olympic cyclists. Seven months before the Beijing Olympics, Anna had a major fall during one of her rides and came as close as two millimeters to being dependent on a breathalyzer. She suffered a fractured second vertebra near her skull, torn ligaments and tendons, a dislocated right shoulder, and skin abrasions. Somehow, through amazing strength and who knows what, she still qualified for the Beijing Olympics, where she won a silver medal.[8] This seemed to be a story of mind over matter. But mind over matter wasn't enough of an answer for me.

What was going on behind the scenes? What was the mind doing that caused these miraculous healings? What did these individuals tap into that changed their biology or circumstances?

Is the Law of Attraction a Fantasy or a Reality?

In 2005, I watched a movie called *The Secret*.[9] It was a revolutionary film that seemed to take the world by storm at the time. A lot of people I knew were talking about it. Some responded positively, and some called it airy-fairy mush. In the movie, they talked about the law of attraction. Put simply, it says that if you ask for something, believe in it with all your might, and are open to receiving it, you will get it—in fact, you can have anything you want. Just as gravity is a law, the law of attraction exists with equal credibility. With this in mind, you can bring wealth into your life, attract relationships, and even become healthy if you happen to be sick.

As I mentioned above, I noticed two main reactions to this idea. On one side were the full believers. They believed it completely and absorbed everything they could about it. They bought the book and changed their lives accordingly. Then there were the disbelievers. They mocked it with sarcastic comments like, "So, if I ask for a diamond ring, it will just appear—yeah, right! Then everyone would be rich. There isn't enough of everything to go around."

I had a different reaction. I could see the possibility that there might be some truth behind the idea, but I wanted to know more. It did seem a little unlikely that this secret was as easy as just asking for something and then having it appear. The movie also touched on a bit of science, so I wanted to find out more. I wanted to know if this could be real, and if so, how it really worked.

The issue of mind over matter poses a lot of questions for me. Interestingly, though, they can all be answered with one theory. After just a few more pages, I will tell you what it is. There's just one more set of questions—probably one of the bigger ones, too: religion versus science.

Questions on mind over matter I will answer throughout this book include the following:

1. How can our beliefs change our realities—or even be a matter of life and death?

2. What can we learn from autism about the possibilities of the mind?
3. How do miraculous healings work, and why don't they work for everyone?
4. Is the law of attraction real, and if so, how does it work?

RELIGION VERSUS SCIENCE

Gods, Beliefs, Faith, and Prayer

As I illustrated at the beginning of this chapter, the question of religion versus science has mystified me since I first learned about the two concepts. As a child, I was never baptized. My parents decided to allow me to make an educated decision when I grew up as to what religion, if any, I chose to follow. As I traveled and went to different schools, I was introduced to different religions. I attended two separate Anglican schools; however, on Christmas Island, the school was a state school, and the religions of the students I met there included Buddhism and Islam. I remember going to Sunday school a couple of times, but I really wasn't too interested. I do remember thinking that it seemed strange that we all came from two people, Adam and Eve. Also, how could God just make the world out of thin air in six days? These were just some of the thoughts I had as a child, but I never really questioned these issues out loud.

Then, at school during biology class, we learned about Darwin's Theory of Evolution and natural selection. We also learned about the Big Bang Theory, that we all came from one point that created everything by exploding and expanding over time.

For me, to see both sides of this story was great, but how could both theories be true? You couldn't really believe both. For one to be true, the other had to be false, and vice versa. I pondered all the religions I'd encountered. Although vastly different in many ways, they seemed to have a few noticeable similarities:

- First, they agreed that there was a **God**. Whether external from us or within us or both, there was a being that had control or influence over our fate.
- Second, there were **beliefs** that were to be followed according to each religion. Some religions had beliefs in such things as heaven and hell. Other religious beliefs included the notion that women's bodies should not be seen in public, that it was a sin for anything other than their eyes to be seen. These beliefs formed the reality of each religion.

- Third, one of the most important similarities was the need for **faith**—faith in the religion's God, faith in its beliefs, and faith in the overall religion. This was of the utmost importance.
- Finally, another similarity I found in all religions was **prayer**. All religions had some form of prayer, whether it was mantras, meditation, or literally addressing oneself to God. Prayer was direct communication, straight from the individual to the Almighty.

I thought it was interesting that all these religions had similarities. Could this just be a coincidence? It seemed there must be something to religion, considering how many types existed. Millions of people practice their religious beliefs every day, so science can't be completely right. Religion must have some truths. Then again, science was pretty good at finding truths—couldn't both be right? I had to find the answers!

Superstitions and Murphy's Law

Superstitions to me are quirky beliefs that don't seem to have logical merit. For example, there is the superstition that if you walk under a ladder you will have bad luck. Then there are others, like the one about the black cat crossing your path, or opening an umbrella inside, or even the idea that breaking a mirror could give you seven years of bad luck. Where did these strange notions come from? And why do chain letters—or nowadays chain e-mails—have such power over people, so much so that they don't want to break the chain in case they have bad luck?

Different cultures seem to have different superstitions. For some, red is a symbol of good luck, as in China. But then I have also heard that it can be bad luck in other situations. In Western culture, the superstition about the number 13 is so ingrained that any reasonably old hotel won't have a room 13 or even a thirteenth floor. Have you ever said, "That won't happen to me—touch wood," and then reached for the nearest wooden thing around you? Whether people believe these things or not, they don't like to take their chances.

I feel Murphy's Law is another version of superstition. It is the belief that the opposite of what you really want will happen. If you really want it to be a beautiful day, Murphy's Law will ensure that it is rainy. However, if you decide to take an umbrella with you, Murphy's Law will almost guarantee that you will get a great day since you have prepared for the worst. How did such a law come about? How did it gain the status of law—so much so that it sometimes really does feel like something is going on? Is there something behind this, or do we just have a tendency to read too much into things?

Questions on religion versus science I will answer throughout this book include the following:

1. What about the dichotomy between science and religion?
2. Why is religion so powerful and important in people's lives, and is there a right or wrong one?
3. Are superstitions real?
4. How does Murphy's Law work?

ARE YOU READY?

With all these questions in mind, my brain was in overdrive. I needed to find some answers, and not only that, I needed to find out what I believed in.

I started doing some research in 2007. During my four years of reading different books, watching documentaries, and attending many courses and seminars, I started to feel that I was formulating consistent answers to some of my questions. The answers to the different individual questions seemed to have similar elements. I was putting everything into place. The paranormal started making sense, unexplained connections were being explained, and I finally felt like I knew where I belonged in the world and what I personally believed in.

If you are ready to step into a new world of answers and possibilities, then feel free to turn the page. Be sure to keep an open mind to the end, and then make your own decision. The journey may cause you anguish, since changes in one's reality can often be scary. It has taken me four crazy years of trying to stretch and open my mind, but the journey has been worth it, and I will never look back.

Buckle up! Your world might turn a little upside down after you open your mind to the possibilities on the next few pages.

If you are ready, turn the page. See you on the other side.

Chapter 2: An Answer

MY THEORY OF EVERYTHING

I came to this theory after a few years of research in metaphysics. As I progressed with my research, I felt I needed to try to decipher the world of quantum physics to truly understand what was going on. Overwhelmed at first, I began to detect a theme that seemed to answer the questions I had buzzing around my brain.

I found the answer after breaking everything down to the smallest particle known to exist and then understanding the characteristics and behaviors of this particle. By comprehending this, I believe I now understand why and how things happen to us as human beings.

In this section, I want to lay these ideas and findings out for you as simply as I can. I have broken them down into four parts:

- understanding the basic building block of existence,
- defining a unique characteristic of this building block,
- establishing the bizarre behaviors of this building block, and finally
- determining how we fit into the puzzle.

Before going into it all, I want to give you a brief summary of what quantum physics is all about. Don't stress, as I am not going to start confusing you with big words and jargon—I will just summarize what it is all about so that when I do dive further into this, you have an elementary comprehension of what I am talking about.

Quantum Physics

Those who are not shocked when they first come across quantum theory cannot possibly have understood it.[10]
—Niels Bohr

Quantum physics is the area of science in which physicists are trying to define how the world works using mathematical equations that explain the behaviors of the smallest known subatomic particles and the forces that influence them. What I mean by forces are things like gravity and electromagnetism. When a ball drops to the ground, it has been affected by gravity. When you turn on your TV or lights, you are using the principles of the electromagnetic force. There are two other forces, which are called the strong and weak nuclear forces. There isn't too much of a need here to understand these, but essentially they are the forces that hold particles together, and without them, the particles in your body wouldn't stay bound together—they would be all over the place. But even though they are very important, it is not essential for you to understand them here.

For the past century or more, physicists like Max Planck, Albert Einstein, Niels Bohr, Erwin Schrödinger, David Bohm, Michio Kaku, and Stephen Hawking have been trying to find the theory that brings all the forces of nature together in one simple equation. Theories have been evolving as more and more physicists suggest alternate ideas. Some of the names of these theories are the Standard Model, the Theory of Relativity, Unified Field Theory, Grand Unified Theory, Superstring Theory, and—one of the latest discussed by Stephen Hawking—M-Theory.[11] You really don't need to worry about the details of these, since I will translate the most important concepts. I just want to give you a background understanding of the material I have been sifting through to come up with my answer.

Now you understand where I am coming from with my research, I want to explain it all in my own words. That way, you'll be able to comprehend the craziness that is the quantum world! If you are a quantum physicist, look away—you may cringe at the way I make something so complex and comprehensive quite simple. It's a little like teaching an elementary school kid about high school topics. I'll be using simple vocabulary to explain phenomena that need a whole new language to be really understood. I may not explain the concepts in detail, but I will give you a basic insight into the world of quantum theory.

Part One: Understanding the Basic Building Blocks of Life

I want to take you back to school for a minute. This may have been an area of school you didn't pay much attention to, as you were busy passing notes to your friends in class. So no passing notes now—I'm watching you! Okay, I want to take you back to that one class in school where you learned about atoms. I believe this should be a much more significant lesson in the lives of kids—I mean, the one where we learn about what we are all made of. I think science teachers don't realize the significance of this kind of lesson. Anyway, hopefully I'll be able to show you what I mean. Now, if you remember, your teachers probably taught you that things are made of materials. A chair is made of wood, clothes are made of fabrics, wire can be made of metal, a watermelon is made of water and flesh, and so on. If we break these materials down further, we find that they are made up of molecules.

Molecules consist of different groupings of elements, and each element, in turn, consists of a single type of atom. Originally, more than a hundred years ago, it was thought that the atom was the basic building block of all things. It is very small, so it's understandable that philosophers and scientists originally thought so. If you arranged atoms in a line, you could fit 50,000,000 of them in a single centimeter! Let's take water as an example of this and break it down. Water is the name of the material, H_20 is the molecule, and a single water molecule is made up of two hydrogen atoms and one oxygen atom. See the diagram for an illustration.

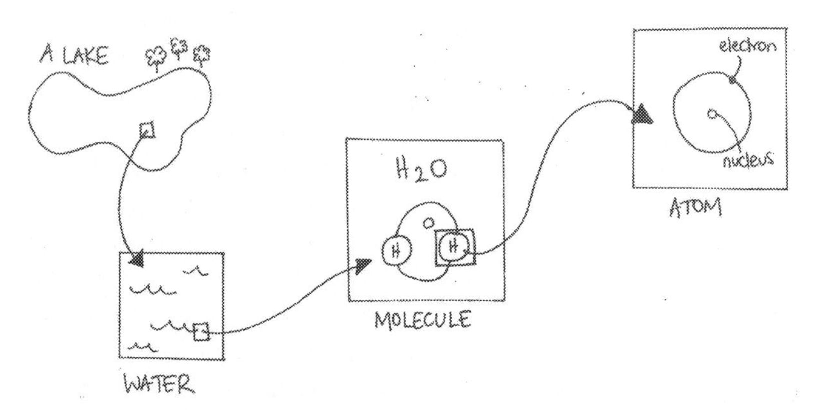

1. The connection between the macroscopic and microscopic worlds: materials, molecules, and atoms.

Okay, I hope I haven't lost you yet. So what makes up an atom? An atom consists of particles known as electrons that whiz around a nucleus. This nucleus contains protons and neutrons. When the number of each of these particles inside the atom changes, the structure of the atom changes, turning it into a different element. For example, the most basic type is the hydrogen atom, which consists of one proton and one electron. I won't go into it, but if you remember back to those periodic tables in chemistry class with copper, zinc, iron, and so on, that's what they were going on about—all the different elements! To this day, all I remember from that table is one of the words from the sentence I made up to help me remember the order of the elements. That word was *hippopotamus!* Not really useful now.

It's at this stage that I want you to understand the makeup of the atom. Like I said, this basic building block of life consists of an electron or electrons whizzing around a nucleus. Your teachers may also have told you that there is a vacuum or empty space between the electrons and the nucleus. Let me give you a real understanding of just how much of this space there is.

Imagine that the nucleus of the atom is the size of a grape. Now place that grape in the middle of a massive, empty field. Let's say that a hundred meters from this grape is a piece of salt. This salt is the electron, and it travels around the grape in a circular motion, always remaining a hundred meters from the grape. What I want you to focus on is that everything between the grape and that tiny bit of salt appears to be empty! I want you to think about this. The basic building block of life seems to be 99 percent empty.

Dr. Bruce Lipton, a distinguished biologist, said, "If you were to get a camera, shrink [yourself] into something smaller than matter, then go through one side of the atom, [and] then take photos right through to the other side—it would be blank."[12]

The actual matter of the atom, the nucleus and electrons, makes up less than 1 percent of the atom's total volume. Let's think about this on a larger scale. If atoms are 99 percent empty, then molecules are 99 percent empty. That means that the materials that make up our world are actually 99 percent empty. That means that you and I are also 99 percent empty. Less than 1 percent of you is actual matter! In fact, William Tiller, Ph.D., says in the documentary *What the Bleep!? Down the Rabbit Hole* (also known as *What the Bleep Do We Know!?*), "You could actually fit the entire matter of the universe into a tablespoon!"[13] How's that for mind-blowing!

This is very hard to accept, as, for example, when I am typing on this computer, it feels quite solid! It feels quite full of stuff. When I look at a billiard ball, it looks quite solid. If it hits you in the head, it definitely feels solid! From my research, I have now learned to accept that what I see on the face of things isn't always the case. Have you ever held a flashlight up to the end of your finger and seen the light penetrate it? You can see the light coming through your finger. We are not as solid as we think we are.

So what science has proven is that everything is made of atoms and that 99 percent of what makes up these atoms is emptiness. It is this empty space that has quantum physicists busy today—and it's also what I'll be delving into in the following pages.

I now want to take you further down on the microscopic scale into this empty space, and it's through quantum physics that we're going to do this. Let's just say that you and I live on a macroscopic scale; that is, we experience the universe in terms of what we can see and measure. All the matter that we can see is on a macro scale. The book, the pillow, the water in your glass—all of it is experienced by us on this macroscopic scale. The microscopic scale of things includes the world of atoms and the particles that make up atoms (subatomic particles). We are going to go even smaller than the particles we've talked about so far: we are going to talk about the much tinier subatomic particles that make up this universe.

So what exactly is that small? What have quantum physicists been talking about, researching, and studying? What is in this nothingness, in this tiny space, that has NASA and British Aerospace so interested? Why has CERN, the European Organization for Nuclear Research, spent over $20 billion on a machine that stretches underground across Switzerland? What has the Institute of Noetic Sciences discovered in this nothingness that has so profoundly changed our way of thinking?[14] What can be so physically insignificant yet at the same time so vastly significant?

Physicists soon realized that the empty space inside atoms can no longer be thought of as truly empty. Max Planck, the physicist known as the founder of quantum theory, performed an experiment in 1911 that helped physicists understand that this "empty space was bursting with activity."[15] Planck realized that everything was made up of infinitesimal packets of energy called *quanta*. It's these quanta or packets of energy that are the basic fabric of the universe. Now get ready to wrap your head around this: he concluded that the actual length of a Planck unit*—its official size—was $1.616252(81) \times 10^{-35}$ meters!

Another way to look at it is that in just over 1.6 millimeters, there are 10^{32} Planck units. That is a one with thirty-two zeros after it. As Stephen Hawking and coauthor physicist Leonard Mlodinow put it in their book *The Grand Design: New Answers to the Ultimate Questions of Life*, a Planck unit is "a billion-trillion-trillionth of a centimeter."[16] Look at the ruler on the next page and realize that more than a billion times a billion times a billion Planck units are in that space! Are you getting how small that is?

---

* Planck length

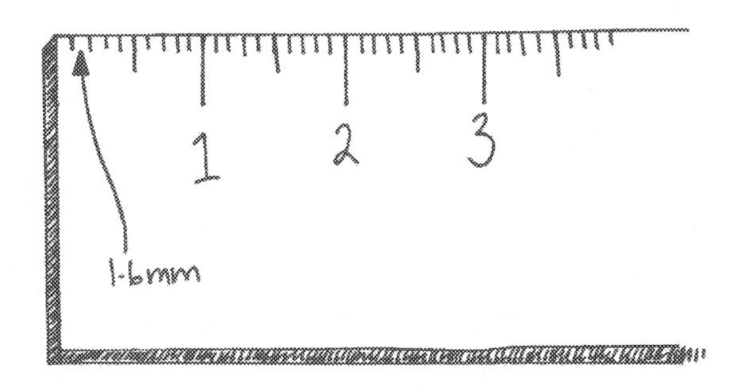

2. Illustration of the amount of space needed to fit billions of Planck units.

So this supposedly empty space I have been going on about is actually full of extremely tiny particles of energy. David Bohm, one of the twentieth century's prominent quantum physicists, said in his book *Wholeness and the Implicate Order*, "What is implied ... is that what we call empty space contains an immense background of energy, and that matter as we know it is a small, 'quantized' wavelike excitation on top of this background, rather like a tiny ripple on a vast sea."[17] Lynne McTaggart talks about this sea of energy in her book *The Field: The Quest for the Secret Force of the Universe*. She calls it the "Zero Point Field—an ocean of microscopic vibrations in the space between things. If the Zero Point Field were included in our conception of the most fundamental nature of matter, ... the very underpinning of our universe [would be seen as] a heaving sea of energy."[18] Penney Peirce puts it this way in her book *Frequency: The Power of Personal Vibration:* "Albert Einstein gave us a great truth: $E = mc^2$ ... the idea that mass or matter could really be very slow, compacted, stored energy, that matter and energy were versions of each other ... A rock is energy, a cookie is energy, a log in the fireplace is energy, the body is energy, I am energy! Everything is moving at different speeds, and matter might convert into other more or less active forms of energy—water to steam or ice, for instance. So what might I convert into?"[19] We are not solid. Our bodies are made up of more than a gazillion vibrating balls of energy. Even the space between you and the book you are reading isn't empty—it's made up of the same thing you are.

To really be able to grasp this, I want you to think about sand glass bottles—you know, the ones you buy near beaches where they've colored the sand and either made a picture or a pattern with it that fills up the entire bottle. Well, I want you to look around you right now. Whether you are in your room, on a bus, at the beach, in a car, or hiding in a closet, imagine that everything in the area around you is made up of about a gazillion bits of sand. So the book or the iPad is made of sand, you are made of sand, even the space between everything is made of sand. I think you get it: you are in a sand glass bottle using sand to create a 3-D picture of the scene you're in.

Now, take this image and adjust it slightly. Change those sand particles into tiny vibrating energy particles. Stop and look around for a second. Imagine it right now. Everything is a mass of tiny vibrating energy particles. You are in an ocean of energy—even you yourself are a sea of energy within the ocean. Pretty interesting, huh? If that has you weirded out, I want you to ponder this for a bit. Fish swimming in water: do you think they actually see the water between them and the other fish, or is it exactly like the air we breathe—which we see as empty space? Perhaps, just as we see the water between the fish, there is something out there that can see the energy between one part of us or our world and another.

These energy particles, though miniscule, are probably the most powerful bits of energy that you can find. Physicists hypothesize that if we can learn to fully utilize these energy particles, we may be able to travel at the speed of light. A great example that shows just how much energy exists inside each atom is given by physicist William Tiller, a pioneer in the field of psychoenergetics, the study of the effects of consciousness on subtle energies. He says, "If we were to take a single hydrogen atom, the energy particles in that vacuum have more energy than a trillion times [what] there is in all the mass of all the stars and planets out to a billion."[20] That is in just one atom! Remember I said there are about 50 million atoms in the length of a single centimeter. I guess you can see why scientists are so fascinated. If we could harness this type of energy, so many things would be possible: traveling at the speed of light, endless energy for generations to come, and so many things that will seem unfathomable to us until we can get our heads around this idea! As Bohm said, "it may be said that space, which has so much energy, is *full* rather than empty."[21]

So scientists have discovered this ocean of energy, but has anyone experienced it directly? Great question! Dr. Jill Bolte Taylor had this fortunate experience under not-so-fortunate circumstances. Taylor, a brain scientist, experienced a stroke during which she suddenly was able to observe and grasp our connection to this universal energy field. She explains her experience from a scientific perspective in her book *My Stroke of Insight:*

A Brain Scientist's Personal Journey[22] as well as on the TED website (www.ted.com), a nonprofit website dedicated to spreading worthwhile ideas.

During the stroke, she was conscious of what she was going through and tried to take notes for scientific purposes. She realized that when the left side of the brain was functioning, she could perform practical mental tasks, such as working out the numbers needed to call for help or the process involved in dialing the phone. However, when her left brain chatter went silent, she experienced something completely different. She describes it like this in her speech on TED: "I looked down at my arm, and I realized that I can no longer define the boundaries of my body. I can't define where I begin and where I end, because the atoms and molecules of my arm blended with the atoms and molecules of the wall. And all I could detect was this energy and in that moment, my brain chatter, my left-hemisphere brain chatter, went totally silent—just like someone took a remote control and pushed the mute button. Total silence. And at first, I was shocked to find myself inside of a silent mind; but then I was immediately captivated by the magnificence of the energy around me. And because I could no longer identify the boundaries of my body, I felt enormous and expansive. I felt at one with all the energy that was, and it was beautiful there."[23] She was interconnected with the ocean of energy.

Let me give you an example of when you may have felt the evidence of this ocean of energy. Remember a time when you walked into a concert or sports match midway and got a shiver of excitement? You just walked into a sea of vibrating high-energy particles, all resonating with the energy and excitement of everyone at that concert. That sense of excitement was palpable!

So what about these packets of energy? What makes them so special? Everything is made of this field of energy. You are 99 percent vibrating packets of energy and only 1 percent matter. But these energy particles have a unique characteristic that can change the way you look at yourself—and even the universe.

Part Two: A Unique Characteristic—Frequency

By now you may be wondering what all this has to do with a tuning fork. Considering a tuning fork is mentioned in the title of the book, one would assume that it's at least slightly significant. Well, guess what—it is! I'll be talking about it in this part of the chapter.

These infinitesimal particles of energy I've been discussing vibrate at many different speeds, known as frequencies. When an energy particle vibrates back and forth, the distance between this back and forth movement is its wavelength, and the number of times this wave occurs per second is its frequency, measured in hertz (cycles per second). The frequency of the energy particle actually determines what it is. For example, one type of high-frequency particle is the X-ray, which vibrates at around the 10^{18} Hz (that is, a vibrating speed of 10,000,000,000,000,000,000 times a second). Slower-moving particles include visible light waves (around 10^{13} to 10^{15} Hz) or the even slower radio waves. Check out the diagram below.[24]

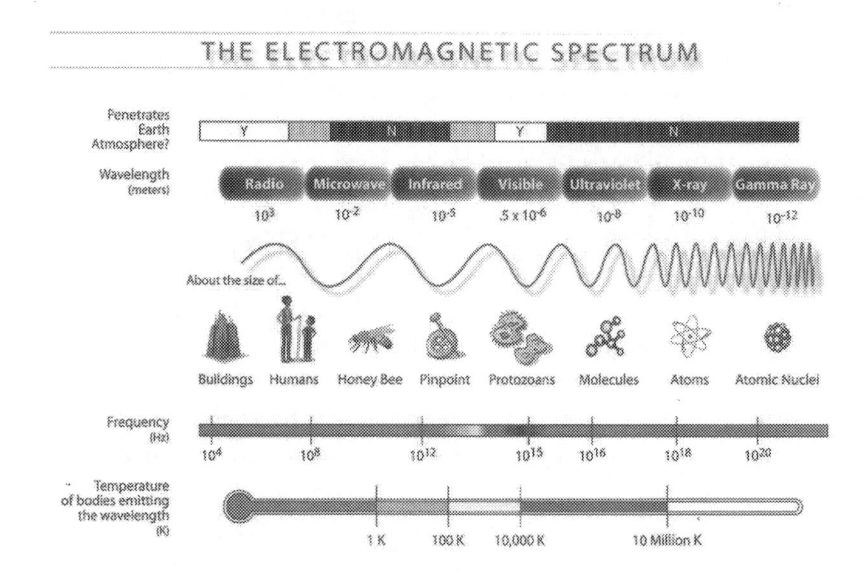

3. NASA's diagram of the electromagnetic spectrum.

So when you tune into a radio station, say 106.9, that particular energy wave is vibrating at 106.9 MHz,* or 106,900,000 times a second. What's happening is that a radio makes its receiver vibrate at the same frequency so that it can read the waves, messages, or music coming in on that wave frequency. Cool—hope I haven't lost you yet!

As I mentioned above, every particle in the universe has an energetic rate of vibration, or frequency. Dr. Richard Gerber says the following in his book *Vibrational Medicine for the 21st Century* (republished as

* Megahertz, or 106 Hz.

A Practical Guide to Vibrational Medicine): "Modern physics tells us that the only difference between these forms of energy is that each oscillates at a different frequency or rate of vibration."[25] Have a look at the diagram on the previous page. The slower the frequency, the more it behaves like a sound wave. Then you have light and color—or what we can see. As the well-known physicist Michio Kaku explains in his book *Hyperspace: A Scientific Odyssey Through Parallel Universes, Time Warps, and the 10th Dimension,* according to String Theory (which I'll explain in the next section), "matter is nothing but the harmonies created by this vibrating string [energy particles]. Since there are an infinite number of harmonies that can be composed ... there are an infinite number of forms of matter that can be constructed out of vibrating strings."[26]

Neurosurgeon Karl Pribram took a different approach to the study of frequencies. He went on an experimental path to discover the location of memory in the brain. McTaggart writes about his work in *The Field*: "His laboratory was the first to discover the location of cognitive processes, emotions and motivation."[27] He studied the brainwaves of animals in many different experiments. Over a few years and after a series of events that included meeting Dennis Gabor (Nobel Prize winner in physics for holography), he came to the following conclusion about our brains: "When we observe the world ... we do so on a much deeper level than the sticks-and-stones world 'out there.' Our brain primarily talks to itself and to the rest of the body not with words or images, or even bits or chemical impulses, but in the language of wave interference: the language of phase, amplitude, and frequency—the 'spectral domain.' We perceive an object by 'resonating' with it, getting 'in synch' with it. To know the world is literally to be on its wavelength."[28]

Ponder that for a bit. When you look at an object, say a couch, your brain isn't seeing a picture of a couch—it's picking up the frequency of the couch. Your brain then interprets it into the picture that you see. Everything is merely vibrating frequencies—not solid matter, but a range of frequencies. It's a little crazy when you really think about it, isn't it?

Our human senses are designed to pick up only a limited range of frequencies. Our ears, for example, are designed to pick up a given range of sound frequencies. As you may know, dogs can hear higher frequencies than we can—our range of audible frequencies is quite limited. Our eyes are similarly designed to pick up a certain range of frequencies, though some people can pick up more than others, and I will talk about this later in the book. Our sense of touch can feel frequencies like heat and cold, our ability to smell picks up another type of frequency, and our taste buds yet another. We are just like radio receivers capable of picking up five different ranges of frequencies. For many of us, any frequencies outside this range are not only invisible but go completely undetected and are therefore assumed not to exist. We have designed some receivers to assist us with our limited receptive ability, such as microscopes, microphones, and speakers, which magnify the frequencies enough that we can detect them. Again, though, we're relatively limited—and why is that? There is an unlimited range of frequencies; we only get a snippet of what is going on in the universe. It's as though we're watching a play through curtains that are only a few inches open, and we see and make our assumptions based solely on this snippet that we can sense. Keep this in mind, because I'll be talking about it a bit more later.

Resonance

Now I'd like to use the example of a tuning fork. A tuning fork is a two-pronged metal fork that is usually used to tune instruments. There is a picture of one on the cover of this book. Its length and size determine its frequency or note. When you tap it, it vibrates at its specific frequency, and then you can match your instrument to this tone.

There is something very intriguing about tuning forks that I think is very important. It is the phenomenon of *resonance*. Let me explain. If you have several tuning forks with various frequencies in one room, and if you pick up one of them and tap it, all the other tuning forks with the same frequency will start to vibrate, hum, or resonate. This resonance produces an experience of intense harmony. All the other forks—those not tuned to this frequency—remain silent, since they can't detect or register the frequency that is resonating around the room. They are unable to sense it and therefore unable to respond to it.

4. An illustration of the tuning fork analogy.

Why do I think that this is an amazing and important phenomenon? Well, we are like tuning forks in that each of us has his or her own frequency—many frequencies, in fact, that together make up each person's own individual frequency. It's almost the same as our fingerprints, although unlike a fingerprint, our frequencies can change. Penney Peirce tell us, "Your personal vibration is the overall vibration that radiates from you in any given moment. That frequency is a combination of any of the various contracted or expanded states of your body, emotions, and thoughts ... It naturally fluctuates."[29] Remember when I explained how many energy particles can fit in just over 1.6 millimeters? That's more than a billion times a billion times a billion. Imagine how many of those particles it takes to make up a human being. Because of the complexities of humans, we have many different frequencies going on at once. Our livers, our hearts, and even our emotions and thoughts give off frequencies. This group of frequencies forms our own personal frequency, which is constantly shifting. As Peirce says, "The vibration of one aspect of your makeup affects the vibrations of the other aspects. If you have been depressed for a while, the low emotional state may cause your imagination to dry up and your thoughts to be more hopeless and negative."[30] That is why you feel amazingly radiant on some days, while on other days you feel down, confused, and chaotic. The frequencies of the energy particles that are *you* are vibrating in different ways.

Another way of looking at this is to compare yourself to an orchestra (hang in there with me!). Each note played by each individual instrument produces one frequency, like a tuning fork. When you combine the notes produced by one instrument, you create a melody that again has its own set of frequencies. When you combine multiple melodies from all the different instruments, you get a beautiful piece of music, which has a massively complex and continuously changing frequency. That is what a human being is like. Just as a piece of orchestral music can have many different emotions and colors throughout, we possess an ever-changing range of frequencies.

We can actually experience this resonance ourselves on many different levels. For example, have you ever sung or heard someone singing and been able to tell whether they were on pitch or not? That kind of intuition is a clue to your understanding of resonance. When we resonate with something, we experience a sense of harmony, a more pleasant feeling than otherwise. Resonance is a feeling that we are naturally attracted to.

Another form of resonance takes place when you have a conversation with someone and you both feel like you understand what the other is saying. You bounce off each other and walk away almost buzzing. The ideas and thoughts you were expressing had frequencies that resonated with each other. Remember that we are made up of a massive number of energy particles that vibrate within a certain range of frequencies. It's these energy particles that I'm speaking of when I talk about your frequencies. It's not something airy fairy—it's the same thing that I illustrated scientifically in the first chapter. Resonance is the key to helping you find your true meaning, or why you are here—something I'll talk about a lot in chapter 4.

Multidimensional Universe

All right, if your mind hasn't been stretched enough, get out the rubber bands, because I am about to flip it into another dimension—literally!

Remember how I mentioned that, as human beings, most of us can only sense a limited range of frequencies—like looking through a narrow gap in the curtain of this infinite universe? Well, what quantum physicists have discovered through working on their many different theories of everything is that these tiny vibrating particles actually vibrate through many different dimensions; that there are more than the four dimensions we know of so far (up, down, sideways, and time); and that we actually live in a multidimensional universe that includes up to or maybe even more than eleven dimensions. For those interested in the science, I will briefly explain it from two different points of view: the scientific and the metaphysical. For those who just want a basic understanding, I will give my interpretation and translation of the scientific jargon. If you find yourself overwhelmed while reading about the scientific background to these ideas, just flip over the page and find the paragraph headed "To Translate."

For some time now, physicists have been on a quest to find a theory of everything, an elegant solution that would explain and unify all the known forces and phenomena in the universe. One such model, called *supersymmetry*, prompted Abdus Salam—a renowned twentieth-century physicist and one of three winners of the 1979 Nobel Prize in physics—to describe it as "the ultimate proposal for a complete unification of all particles." Scientists are still working out the details of these theories and making new discoveries in the process.

Here are some brief descriptions of the models they have used, followed by a statement about an alternative branch of study, metaphysics, which is being found to overlap with quantum physics and other areas of science in remarkable ways.

There are four known fundamental forces in nature (a force being that which affects matter):[31]

1. electromagnetism
2. gravity
3. the strong nuclear force
4. the weak nuclear force

Quantum Field Theories are mathematical equations that explain the behavior of each of these four forces individually.

A *Grand Unified Theory*[32] tries to combine three of the field theories (electromagnetic, strong nuclear, and weak nuclear) into one mathematical equation. GUTs, as they're called, are seen as the next step to TOEs (theories of everything), because they do not yet include gravity as part of the unified equation. This inability to discover an equation that includes gravity is what has left quantum physicists unsatisfied with this class of theory.

String Theory is a model that seems to unify all four forces. However, it hasn't provided a sufficiently beautiful, or symmetrical, equation, so some quantum physicists remain unsatisfied with it. What string theory did discover and break through was the following, as described by Hawking and Mlodinow in *The Grand Design*: "According to string theory, particles are not points but patterns of vibration that have length but no height or width—like infinitely thin pieces of string. String theories also lead to infinities, but it is believed that in the right version they will all cancel out.

They have another unusual feature: They are consistent only if space-time has ten dimensions, instead of the usual four."[33]

M-Theory is a solution that string theorists came up with as yet another fundamental model that seemed closer to their goal of a symmetrical equation that would explain how everything worked together in the universe (no one really knows for sure what the *M* stands for). Hawking and Mlodinow say that "M-Theory has eleven space-time dimensions, not ten" as originally thought by string theorists. They are time, the three macroscopic spatial dimensions that we commonly experience, and seven curled dimensions (*curled* meaning that they are rolled up into a miniscule, invisible space) of what M-theorists call *internal space*, which is inaccessible to our senses. Hawking and Mlodinow continue, "String theorists had long suspected that the prediction of ten dimensions might have to be adjusted, and recent work showed that one dimension had indeed been overlooked. Also, M-Theory can contain not only vibrating strings but also point particles, two-dimensional membranes, three-dimensional blobs, and other objects that are more difficult to picture and occupy even more dimensions of space, up to nine. These objects are called 'p-branes' (where 'p' runs from zero to nine)."[34]

The International Academy of Consciousness, a nonprofit organization researching conscientiology (the science of consciousness) has found through both subjective and objective studies that we live in a multidimensional universe. Consciousness, which I will discuss more in part 4, can and does move across all dimensions. They have found that the different dimensions are all in the same space and time, just at different frequencies. The higher the frequency, the more subtle the energy and dimension. Our physical bodies are restricted to the frequencies they can perceive, whereas consciousness is more flexible in this area.[35]

To Translate All right, if I lost you up there, that's okay! It has taken me a few years to wrap my head around all this. The first time I even approached these concepts, my brain shut down. It simply turned off, stopped paying attention, and concentrated on something far less brain-consuming, like what to eat for dinner.

Let's go back to the start, and I'll explain it in simpler terms. We are made up of all those miniscule vibrating particles of energy. Quantum physicists have found that these particles that we would logically assume are little round balls, so to speak, are actually tiny bits of flat string (like ribbons). Not only do these bits of string vibrate at different frequencies, some of them can change shape by connecting their ends together to make a circle or split in two to make two bits of string, and so on.

What this means to us laymen is that these stringlike packets of energy have an infinite range of vibration. There is no known highest frequency for them; nor is there a lowest. We, as human beings, can sense only that small gap in the opening of the curtain onto an infinite stage. The other levels to each side of us are even more dimensions that we can't yet sense.

Let me try to use a visual example. Imagine a lion standing behind the curtains that are only open a few inches so that all we can see from our vantage point or dimension is his tail flickering. We wouldn't know what was causing this tail to flicker, let alone that it was even a tail. All we could see would be a cylindrical furry thing wriggling around with some fluff coming out of nowhere at its tip. If our senses could detect more than the four dimensions that we currently experience, we might be able to see that this furry thing is actually part of something much bigger—a lion. This is just an illustration of a principle that can show us how multidimensional thinking might shed some light on real examples of unexplained phenomena. Some of these unexplained phenomena may actually make sense when you look across more than just a few dimensions.

Now I want to adjust this analogy ever so slightly. I want to eliminate the confusion of time and space. In the first example I used a stage that went on forever on either side of our dimensional experience. However, that isn't quite what things are like here. The dimensions aren't on either side of us—rather, they all occupy the same space. Think of a house. Within the house are radio waves, infrared rays, microwaves—all sorts of vibrations all occupying the same space. They are just at frequencies that we, as humans, don't register. Well, the same is true for the infinite number of other frequencies that exist. They all occupy the same space that we do—we just don't pick up on them since they aren't in our range of perceptible frequencies.

Let's try using the example of a telescope that allows you to see all the different frequency levels. Imagine that instead of changing its focus, you could tune it to different frequencies (like a radio), and that this would change what became visible to you. The more you turned the knob, the more frequencies you would be able to see. Let's imagine now that you're looking into someone's living room at our normal range of frequencies through this unique telescope. Okay, now we are going to turn the knob to another frequency that's normally inaccessible to us. The forms that you saw in our usual dimension, such as the couch, the TV, or human beings, might disappear. Other frequencies, even other consciousnesses, shapes, and forms might manifest themselves. What I am saying is that in the same space now occupied by the living room, there could actually be something else at a different frequency. Crazy, huh? In other words, there are multiple dimensions all around you and even within you; it's just that we are completely oblivious to them because of the limited receptive abilities of our five senses.

Now, you may be saying to yourself, "All this stuff is just based on theories, not fact. Why should I even pay attention to it?" Well, I thought that myself, but what I didn't realize was that a lot of the technology we use today is built on the basis of these theories. So fact or not, if the theories are able to produce working technology, that shows me that they are on the right track. Michio Kaku put it well when he said that "if quantum mechanics were incorrect, then all of electronics, including television sets, computers, radios, stereo, and so on, would cease to function. (In fact, if quantum theory were incorrect, the atoms in our bodies would collapse, and we would instantly disintegrate ... Thus the fact that we exist is living proof of the correctness of quantum mechanics)."[36]

Just to recap, everything is made of tiny, string-like energy particles that vibrate at different frequencies. Within each frequency is a resonance that could potentially attract or strike a chord with us, depending on our own frequency. The range of frequencies at which the particles can vibrate is unlimited, which suggests multidimensionality. If you have grasped these basic concepts, then you are doing well! If not, just hold on—because in chapter 3, I will eventually be going through the science and answering all the questions from chapter 1. With the examples I give, it all might start to make sense. For now, however, I want you to learn that not only do these particles have an unlimited range of frequencies, they have been shown to behave in a completely different way from what we've come to expect in the macroscopic world we live in. So let's take a look at quantum behaviors!

Part Three: Bizarre Behaviors

The next piece in this brain-stretching puzzle is to realize that energy particles do not behave like the solid reality we have been brought up on—that, until now, there seemed to be restrictions as to what solid things could and couldn't do. Concepts that we assumed to be normal, such as one solid object being completely separate from another solid object, the impossibility of an object being in two places at once, the inability of objects to go back in time or move forward faster than time itself—these are just some of the common-knowledge laws that we've taken for granted. Well, I am about to completely turn all of that upside down! Not only did these amazing quantum physicists discover that we are not solid—that we are a mass of tiny vibrating energy strands—but they also discovered that on the Planck scale of things, these tiny energy strands behave in their own way. They never behave the way we might expect. So fasten your seatbelts and hold onto your brain, since this is where it starts to get really interesting!

Here are the explanations for some of the quantum theories:

1. Entanglement or Nonlocality

Entanglement theory says that any particles created together are interconnected through space and time. Louisa Gilder explains this in her book *The Age of Entanglement:* "It starts when they interact; in doing so, they lose their separate existence. No matter how far they move apart, if one is tweaked, measured, observed, the other seems to instantly respond, even if the whole world now lies between them."[37] Albert Einstein called entanglement "spooky action-at-a-distance."[38] Since then, physicists around the world have been able to replicate this phenomenon. In his book *Entangled Minds: Extrasensory Experiences in a Quantum Reality,* Dean Radin says, "Today we know that entanglement is not just an abstract theoretical concept, nor is it a quantum hiccup that only appears for infinitesimal instants within the atomic realm. It has been repeatedly demonstrated as fact in physics laboratories around the world since 1972."[39]

In our macro-level world, we generally experience objects as separated by time and space. Take, for example, a marble on one side of the world and another on the other side. In the current understanding, if you poke one marble, the only way it could affect the other is if the two were connected via a radio/electromagnetic wave or physical cord right around the world. Even then, if one was poked, it would take some time for a reaction to occur. This new world of entanglement means that these two marbles, as subatomic particles, will actually instantaneously react in the same way if either one is prodded, regardless of time or space. Another word for entanglement is *nonlocality*. As applied to subatomic particles, it suggests that they are capable of nonlocal interaction (that they can be affected by forces or particles not necessarily in their immediate area). Quantum nonlocality was first demonstrated by Einstein, Podolsky, and Rosen in 1935.[40] It essentially contradicts the way we experience the world at the macroscopic level by saying that two separate particles can be affected by something at the same time, even at a great distance, through entanglement.

What does this mean for us? If time and space are no longer relevant, maybe we can travel through time, since it doesn't really exist, and if we are interconnected, what I do may affect someone across the world. Does this entanglement go beyond the subatomic world? Has it been proven in the macro world? Well, I am glad you asked.

Experiments have been done that show that this definitely occurs on a macro level. French researcher René Peoc'h conducted one such experiment with pairs of rabbits.[41] Some of the rabbits were from the same litter and knew each other from living in the same cage, while others did not know each other and had lived in isolation. All pairs of rabbits were separated, placed in soundproof cages, and then isolated from electromagnetic influences. By monitoring the blood flow through their ears, Peoc'h could tell when the rabbits were experiencing stress. What he found was a statistically significant difference between the reactions of the pairs of rabbits that knew each other and those that didn't. Those that did had an almost simultaneous (within three seconds) telepathic connection: their stress levels showed an almost instantaneous mirroring of the stress level of the rabbit with whom they were paired.

5. Rabbits interacting telepathically.

Another experiment I saw in the documentary *The Quantum Activist* with Amit Goswami illustrated two people communicating nonlocally.[42] The experiment was set up at the University of Mexico. Two people meditated together for twenty minutes and were then separated and placed in two different buildings. Both were placed in individual chambers that were electromagnetically impervious. The researchers then hooked each person up to an electroencephalography (EEG) machine in order to measure the electromagnetic activity of their brain waves. They then flashed a light into the eyes of one of the two people and examined the brain wave pattern. When they compared it to the person's whose eyes had had no lights flashed into them, they found that the patterns were almost identical, implying a telepathic connection.

In his book *Entangled Minds,* which reviews the effect of entanglement on the macroscopic world, Dean Radin relates the following observation by *New Scientist* writer Michael Brooks: "Physicists now believe that entanglement between particles exists everywhere, all the time, and have recently found shocking evidence that it affects the wider, 'macroscopic' world that we inhabit."[43]

As entanglement has shown its intriguing face in our macro world, it has brought some unexpected results to our attention. Not only has it revealed to us that space is no longer as we thought, but it has also given us a new perspective on time. Here is the story of an experiment that really boggles the mind in that respect:

In *The Field,* Lynne McTaggart illustrates some of the results that show that even the macro world does not behave according to our assumptions about time. McTaggart writes about Helmut Schmidt, a scientist who "rewired his REG [Random Event Generator] to connect it to an audio device so that it would randomly set off a click, which would be taped to be heard in a set of headphones by either the left or the right ear. He then turned on his machines and tape-recorded their output, making sure that no one, including himself, was listening. A copy of the master tape was made, again with no one listening, and locked away. Schmidt also intermittently created tapes that were to act as controls ..."[44] About a day later, volunteers would take the tapes home and listen to them while trying to influence the distribution of the clicks. This was exactly what happened—even though it had been recorded before. "Their influence had reached back in time and affected the randomness of the machine *at the time it was first recorded.* They didn't change what *had* happened; they affected what would have happened in the first place."[45] Doesn't that just make the mind implode! We can change our pasts as much as we can change our present. What implications does this have? I'll leave that one for you to contemplate.

One familiar example of entanglement for many of us is the movie *The Matrix.* It illustrates our interconnectedness, how we are entangled with everything. If you remember, at the end of the film, when Neo can suddenly see the matrix, he becomes a part of everything and knows what his enemies are thinking before they do. He is entangled with the matrix—they are one.

This is where I am heading with this: we are all interconnected. I am a part of you as you are a part of me. You are a part of this computer I am typing on as well as of that tree outside my window. This ocean of energy or zero-point field is completely interconnected.

So to sum it up, those tiny little stringy energy particles not only vibrate at different frequencies as well as in and out of different dimensions, but they are all interconnected with one another, and they completely disregard our concepts of time and space. Hope you are keeping up, because we are about to go even farther down this vortex of craziness!

2. *Superposition*

Welcome to Hayley's house of tricks! Well, that's what you may think when you read about this next theory. What is interesting about this theory, though, is that like entanglement, it is being replicated in laboratories around the world—turning the theory into reality.

Superposition essentially states that one of these vibrating energy particles can actually be in more than one place at the same time. Physicist and psychoanalyst Dr. Jeffrey Satinover talks about this phenomenon in an interview called "Entanglement—the other side of the coin," part of the documentary film *What the Bleep!? Down the Rabbit Hole.*[46] He explains that it is possible for these particles to exist simultaneously everywhere and nowhere before you actually observe them. However, the instant you look at one of them, it collapses into one possible location, giving the illusion that it is in one place at one time. I'll talk about this observation phenomenon soon, but what I want you to focus on first is the fact that these vibrating particles can exist in more than one place at the same time. So *superposition* refers to the simultaneous existence of a particle in all possible states or locations before an observation has occurred to collapse the other possibilities into only one possibility. As Penney Peirce writes in her book *Frequency*, "a quantum entity exists in multiple possible realities called *superpositions*. As soon as an observation or measurement is made, the superposition becomes an actual reality ... The many become one."[47]

With this idea in mind, you might be thinking, "Yeah, okay, so there is a possibility of things existing in many places, but when we look at things in our macro world, they exist in only one place at a time, so this phenomenon isn't really that interesting." I can see where you're coming from. What if I told you that the existence of one particle in more than one place at a time has not only been observed but actually photographed—in fact, that a single particle has been photographed in even more than three thousand places at one time?[48] That's right, a picture of *one* particle in not just *two* but *three thousand* different places at the same time. It's not half a particle spread over two places—it's the same one in two (or, in this case, three thousand) different places. When you really think about it, it's a hard concept to grasp.

Again, like entanglement, superposition has been observed in our macro world through numerous experiments. Theoretical physicists Tony Rothman and George Sudershan, both significant contributors to various areas of physics, among other things, wrote in their book *Doubt and Certainty*, "The first such experiment, using a potassium atom as the target, was carried out at the University of Rochester in 1991 by Carlos Stroud and John Yeazell. In 1996, Stroud and Michael Noel showed that one was in truth observing the same electron in two locations. Several other such experiments have been performed, with various techniques, but the basic objective is the same—to observe a quantum superposition at almost macroscopic separations."[49]

So what are the implications of superposition? Ever been invited to two awesome events on the same weekend and couldn't decide which one to go to? Five, ten, or two hundred years down the road, with advances in our understanding of superposition, this problem could be solved! But seriously, I want you to again add this to your new basic understanding of how the universe works and to compile it with everything else I have shown you. If all you take away from this is the realization that there might be more to what you have assumed to be real than meets the eye, then I'll be happy.

3. Quantum Holographs and Morphic Resonance

So what if I told you that these particles were not only entangled and could be in more than one place at a time but that they also contained all the information of the universe? Well, according to science, this is a reality. Dr. Edgar Mitchell, former astronaut and founder of the Institute of Noetic Sciences, explains in his book *The Way of the Explorer*, "Recent research ... suggests that we, and every physical object, have a resonant holographic image associated with our physical existence. It is called a quantum hologram. One can think of this as a halo, or 'light body' made up of tiny quantum emissions from every molecule and cell in the body. In other words, the totality of our physical and subjective experiences can be thought of as a multimedia hologram resonant with ourselves and the zero-point field."[50]

Confused? Let me put it simply. A hologram has the unique characteristic that, even if you cut it up, each part still contains the whole and can make a replica of the hologram. It contains the information for the entire hologram, almost like a strand of DNA in a human being, which has all the information needed to make up the human body. Dr. Mitchell says in the same book, "With science we came to recognize that nature stores information in a number of ways, such as the DNA code. Atomic matter itself, being organized energy, also carries information."[51] A quantum hologram essentially contains all the information of the universe. Now, doesn't that blow your mind? One of those tiny vibrating energy particles has all the data that makes up this universe.

Author and researcher Michael Talbot discusses this thoroughly in his book *The Holographic Universe*, where he combines the holographic theories of neurophysiologist Karl Pribram and physicist David Bohm: "Our brains mathematically construct objective reality by interpreting frequencies that are ultimately projections from another dimension, a deeper order of existence that is beyond both space and time: The brain is a hologram enfolded in a holographic universe."[52]

So each quantum particle contains information about the whole universe, just as a piece of a hologram contains all the information of the whole hologram. Hence the name holographic universe. It doesn't matter which vibrating bit of energy you pick up; the answer to any part of the universe can be found in it. And that is a code that scientists are trying to crack—the information contained in those infinitesimal vibrating energy particles.

If we combine entanglement and holographic theories, we get something similar to what biologist Rupert Sheldrake calls a morphogenetic field. Sheldrake sees morphogenetic fields as spanning across plants, minerals, animals, and human beings, and he believes that the information contained within these fields is constantly updating itself from all data across the world as species evolve. In her book *Hands of Light: A Guide to Healing Through the Human Energy Field*, Barbara Ann Brennan says that Sheldrake "proposes that all systems are regulated not only by known energy and material factors but also by invisible organizing fields ... According to the hypothesis, whenever one member of a species learns a new behavior, the causative field for the species is changed, however slightly. If the behavior is repeated for long enough, its 'morphic resonance' affects the entire species. Sheldrake called this invisible matrix a 'morphogenetic field.'"[53] In simple terms, what this is saying is when an animal learns a behavior, this information is instantly uploaded into the energy web of information, which is then accessible subconsciously to all the other animals of that species.

Essentially, whether it's holographic or morphogenetic, the basic idea remains the same. Everything is entangled within this complex, interconnected vibrating information web that defies time and space. Accessing even a part of this information is potentially mind-blowing and perhaps even reaches into the world of the paranormal.

4. *Wave/Particle Duality*

If your head isn't hurting yet, then grab the paracetamol, because this one is a doozy!

Quantum physicists decided to see how subatomic particles behaved by using what is known as the famous double-slit experiment. "The double-slit experiment was first carried out in 1927 by Clinton Davisson and Lester Germer, experimental physicists at Bell labs who were studying how a beam of electrons ... interacts with a crystal made of nickel. The fact that matter particles such as electrons behave like water waves was the type of startling experiment that inspired quantum physics."[54]

Let me explain how this experiment works by using everyday items as examples. Imagine cutting two vertical slits in a board of wood. Behind the board is a plain white wall. If you shot paint balls through the two slits, what pattern do you think they would make on that back wall? That's right: there would be two colorful lines behind the slits at the area of highest impact.

Now, let's do this with water instead of paint balls. If you put this board with two slits halfway in the water and had the back board at the same distance as before, and if you then made a wave pass through the two slits, you would see waves coming out of the slits behind the board and then crossing over and mixing with the other waves. That is, the waves would collide with each other, canceling parts of each other out and impacting the back board in a way that creates alternating lines of light and dark. This is called an interference pattern.[55] See the diagram on the next page to help you understand.

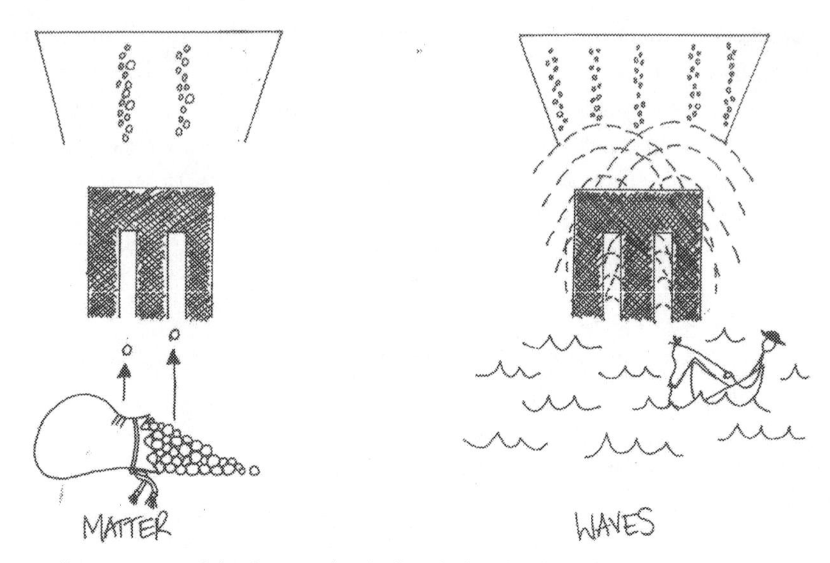

6. Illustration of the basic idea behind the double-slit experiment. The spaces between the lines in the diagram on the right form part of the alternating dark and light pattern.

Let's shrink this whole experiment down to the Planck scale—to the size of the tiny subatomic world—and see what happens. When quantum physicists shot subatomic particles through the two slits, they expected them to act like the paint balls in our example. Instead, what they found was that the board showed the kind of pattern normally left by a wave! This even happened when they fired the subatomic particles one by one. How were they hitting each other to cause the ripple effect? These particles were behaving like waves. How bizarre. They also found that the subatomic particles were actually going through both slits, through neither slit, and through only one of the slits (as in superposition).[56]

Finding these results strange, they decided to observe the particles as they went through the slits to see how it was possible that they were behaving like waves. However, when they did this, the wave-particles changed their behavior and again acted like particles. As the physicists had originally expected, the back board showed a pattern of just two slits, like the one created by the paint balls. It seemed the natural (unobserved) result was for the particles to act like waves; however, when observed, they again behaved according to what physicists at the time expected of a particle.

So not only do you now have to comprehend the concept that we are tiny strands of vibrating energy with an infinite range of frequencies; that these particles, or strands, are interconnected; that they can be in more than one place at a time; and that they contain all the information of the universe; but also that they can behave like either a particle or a wave. That these particles only act like particles when they are observed—what implications does that have for us? Aren't we made of particles? So do we only exist because we observe ourselves? I guess the question "If a tree falls in the woods with no one around, does it make a sound?" will need to be rephrased: "If a tree falls in the woods with no one around, does it actually exist?"

Wow, paradigm-shifting enough for you? Well, this is just the tip of the iceberg. Let's go under the water and see what's behind all this. Let's answer those big questions that you may be wondering about:

- What causes these infinitesimal vibrating packets of energy to decide their behavior? What collapses a wave function and turns it back into a particle? In other words, what causes a nonlocal particle that is in every place at once to focus in on one location?

What is the answer?

The answer is actually simpler than the question itself. The answer is you, the observer.

Part Four: The Observer

The *act of observing* changed the double-slit experiment. Dean Radin says that "this interpretation asserts that consciousness is the fundamental ground state, more primary than matter or energy."[57] This is probably one of the most profound discoveries of quantum physics: we are not separate from our experiments—the very act of observing them can influence them. McTaggart explains it well: "Subatomic particles existed in all possible states until disturbed by us—by observing or measuring—at which point, they'd settle down, at long last, into something real. Our observation—our human consciousness—was utterly central to this process of subatomic flux actually becoming some set thing ..."[58]

Let's break this down. We, our physical bodies, and everything in the space around us, are made up of tiny, vibrating energy particles. They vibrate at an infinite range of frequencies, have unusual behaviors in that they can be in many places at once, are completely entangled, and have the potential to act either as a wave function or a particle. All this is completely influenced by our observation, our consciousness! Wow! Really take that in. This isn't something you should just read and think, "Oh, that's interesting," and then go and flick on the TV as though it were just another regular day. This is completely paradigm shifting and potentially reality changing (literally).

When you walk around, your consciousness affects the mass of particles in the reality around you. You are the center of your reality, and it is your consciousness that has the potential to control this reality. If you let your consciousness be controlled by random thoughts and the limitations you have been brought up on, then you will see your world through those limitations and thoughts. That is pretty profound! All of a sudden, the responsibility is on you, and you become a significant cog in the wheel of your reality.

Researchers Robert Jahn and Brenda Dunne conducted a series of experiments called the Princeton Engineering Anomalies Research (PEAR), one of the most profound research projects illustrating our effect on matter through the use of intention. PEAR tested to see if a person could will a Random Event Generator (REG) to achieve a certain result. "In one 12-year period of nearly 2.5 million trials, it turned out that 52 percent of all the trials were in the intended direction and nearly two-thirds of the ninety-one operators had overall success in influencing the machines the way they'd intended ... So long as the participant willed the machine to register heads or tails, he or she had some influence on it a significant percentage of the time."[59] The act of willing the change actually caused a change in the results, showing again that we, consciousness, affect matter. Profound, really—isn't it? On a lighter note, McTaggart relates that "this might explain all the well-known stories about people having positive or negative effects on machines—why, on some bad days, computers, telephones, and photocopiers malfunction."[60] I have a friend whose computer or phone always seems to go crazy, and the IT guys and phone technicians can't see anything wrong with them. I have always had a gut feeling that my friend was affecting them—and now I know my intuition was spot on. His consciousness, thoughts, and emotions had an effect on his environment.

Consciousness

So what is consciousness? My current primitive understanding is that consciousness, also known as the essence of who we are—our mind, spirit, soul, samadhi[61]—is the key to our reality and existence. It *is* who we are. We are not our physical bodies but the consciousness that resides within. The very awareness of our existence comes from our consciousness. Have you ever been astounded by the way you reacted to something, so that you thought, "That's not me"? Who do you think thought that? The reaction you had was either a physical or emotional response to something: that thought was the essence of you, your consciousness.

The study of consciousness is now the key to everything. Nonprofit organizations like the Institute of Noetic Sciences (IONS)[62] and the International Academy of Consciousness (IAC)[63] are scientific organizations that are trying to decipher just this issue. Quantum physics may have discovered the profound notion that consciousness is the key to affecting reality, but it's now up to our own personal research as well as organizations like IONS and IAC to really discover what consciousness is and how this key to our reality can be utilized for the greater good.

One understanding of the IAC is that consciousness exists across all the dimensions that I mentioned previously—that although our physical bodies are restricted to this dimension or frequency range, our consciousnesses can actually exist across all of them. In other words, consciousness can sense all frequencies, which explains perhaps why some people claim to see different frequencies, such as auras or even ghosts (other-dimensional consciousnesses). Our existence through our physical bodies, according to the current findings of the IAC, is known as *intraphysical*. Experiences in more subtle dimensions outside the dense dimension of this physical reality are called *extraphysical*. I will go into this more later in the book and also discuss how these findings suggest the plausibility of out-of-body experiences, near-death experiences, and perhaps even paranormal phenomena.

The IAC's understanding, through both subjective and objective research, is that consciousness manifests through four types of bodies that together make up the whole of who we are. We have a range of bodies, each at a different frequency, and can therefore exist in different dimensions. The combination of these bodies is known as the *holosoma*.[64] They are: the physical body (*soma*), the energetic body (*holochakra*), the emotional or astral body (*psychosoma*), and the thought body (*mentalsoma*).[65,66] It is through these vehicles that our consciousness can experience different dimensions as each of them vibrates at a different frequency.

But it isn't just the IAC's opinion that there are different bodies. The well-known theoretical quantum physicist Dr. Amit Goswami has come to a similar conclusion and discusses it in his book *Physics of the Soul*.[67] He has deduced that we in fact have five bodies, which he describes beginning with the more ephemeral and ending with the physical: "The outermost is the unlimited bliss body; the next body is theme, or supramental intellect, that sets the contexts of movements of the mental, the vital, and the physical.

Of these latter bodies, the mental gives meaning to vital and physical movements, and the vital has the blueprints of the biological forms of life manifested in the physical. Finally, the physical is the 'hardware' in which representations ('software') are made of the vital body and the mind."[68]

Essentially, both have come to the conclusion that we are not just what we see, that there is more to us on a multidimensional level. Not only are we multidimensional, but our reality is also multidimensional; and it is we—our consciousness—that affects what is going on around us. It is up to us, then, to decide what kind of reality we want and to begin to exemplify it. A fictional movie I saw that really opened my mind to the idea of consciousness is called *What Dreams May Come*. There is a line from that movie that I feel summarizes everything in one sentence: "Thought is real. Physical is the illusion."[69] Another way I like to put it is this:

> "Consciousness is real. Physical reality is its playground." It is up to you to choose whether you want to play on the swings of reality or just sit and watch from the outside.

I'd like to break this process down from its beginning in consciousness to the collapse of the wave function. What I have found through my research is the following:

- Consciousness produces thoughts, which in turn affect our emotions.
- These thoughts and emotions lead to the manipulation of energy vibrations.
- In this way, they change our physical reality.

I call this is the *path of affect:*

Consciousness → Thoughts and Beliefs → Emotions → Energy → Reality

Let me break it down for you.

Thoughts and Beliefs

"There is no way to remove the observer—us—from our sensory processing and through the way we think and reason. Our perception—and hence the observation upon which our theories are based—is not direct, but rather is shaped by a kind of lens, the interpretive structure of our human brains."[70] So what shapes this interpretive lens? It is the thoughts that come from our consciousness. These thoughts, over time, and with experience and knowledge, turn into beliefs that start to structure our reality. Esther and Jerry Hicks say through Abraham that "A belief is just a thought [we] keep thinking."[71] A belief is just a thought that has concretized itself into the fabric our being. It can be as strong as believing that the sun rises every day, which we believe to be an absolute truth. In fact, there is no such thing as absolute truth: nothing is permanent—we are in a state of constant change. One minute we may believe that the world is flat, and then all of a sudden we may be shown otherwise. All beliefs can be changed, just as we often change how we think about things. And this is the key to changing the reality that we see through our consciousness.

Let me give you an example of how people's beliefs can affect physical reality:

"The brain is so good at model building that if people are fitted with glasses that turn the images in their eyes upside down, their brains, after a time, change the model so that they again see things the right way up. If the glasses are then removed, they see the world upside down for a while, then again adapt."[72] Their beliefs rewired their own brains to conform to the reality that they knew was true, which was the direction of the images. If our beliefs can rewire our brains on a biological level, imagine the possibilities of what they could change in our everyday reality on a subatomic level. I once heard a great quote: "We do not see things as they are. We see them as we are."

If it's true that we see our reality as we are, then this puts a greater emphasis on the thoughts we entertain and the beliefs we accept as true.

So what fuels these thoughts and beliefs that in turn affect our reality?

Emotion

Gregg Braden elaborates on thought frequencies in his documentary *The Science of Miracles*.[73] He says that when thought meets emotion, it sends out a feeling. It's this feeling that fuels the intentions of our consciousnesses, which in turn influence our reality. The Institute of Noetic Sciences has recently discovered that the emotion of love does indeed cause the heart to emanate stronger magnetic waves. And it is this emotion of love, with its positive intentions, that affects the reality of the consciousness that is giving or receiving love.

Another very important point that Braden makes has to do with setting one's frequency. Braden researched the prayer forms of different ancient cultures. In one conversation he had with a Tibetan monk, the monk told him, "What you are feeling is actually the presence of what you are asking for actually being there."[74] In other words, it's not our thoughts or words, such as, "Dear God, please bring peace and harmony to the world," that create an effect; rather, it's what we actually do and feel—our sense of the presence of peace as already being there—that produces outward change in the world. Essentially, you need to put yourself in the frequency of already having what you want, not just saying it, but *feeling* it right to the core of those vibrating packets of energy. It's that kind of depth that you need to collapse the wave function (which is another way of saying "the wave of possibilities") into your reality.

Let me use my radio analogy again. In a city or town, there are usually many different radio stations and therefore many different radio waves being sent out at any one time. Now, to get a station that you want, you first have to find the frequency of the radio program that you like. So you turn the dial or push the buttons until you've set your radio to the right frequency, and lo and behold, the frequency matches the radio wave of your program, and you have a connection. All the messages, news, and music associated with that frequency come through.

Now, I want you to think about this in the following way. As you already know, everything on earth—in the universe—is energy, and everything has its own individual frequency. You have a radio inside you: it's called your belief system, and your consciousness can choose the station it wishes to tune into. The frequencies of the different beliefs you entertain in different areas of your life are what shape your experience of the reality around you. How do you think your thoughts are affecting your reality? Are your thoughts helping you through life, or are they holding you back? Stop for a minute and look at what is going on around you right now. Do things always seem to get in the way of your success, or are you a person that always seems to have things work out for the best? Let's explore this a little further.

Energy

Our present consciousness, or state of mind, is the source of our thoughts and beliefs, which in turn are fueled by emotions, both positive and negative. It is these fueled thoughts that end up affecting our reality or, to put it in quantum terms, collapsing the wave function of those tiny vibrating energy particles, which are the threads in the fabric of our world.

One very telling piece of evidence that relates to this notion that thought combined with emotion affects energy and, in turn, reality is illustrated by the experiments of Dr. Masaru Emoto, author of *The Hidden Messages in Water.*[75] Basically, Dr. Emoto labeled water with different messages intended to cause emotional reactions and then froze the water to create water crystals. What he found was that different types of crystals would form, some better than others, depending on the message. Some wouldn't form at all. The crystal with the label "I love you" created a beautiful formation. The one labeled "You make me sick, I want to kill you" barely formed a crystal, and it was messy. The one that was blessed by a monk formed one of the largest, most perfect crystals. Dr. Emoto also subjected the water to various types of music, a natural vehicle for emotion as expressed through musical tones. The results were interesting. The water sample that experienced Mozart's music created an exquisite crystalline form. The abomination that barely formed had experienced heavy metal music. These experiments show the ability of thoughts and emotions to influence energy particles and therefore matter.

Wait, there's a catch!

Unfortunately, there is a catch to this new paradigm. If your consciousness chooses to collapse the possibility that everything I have just talked about doesn't exist, that is exactly what your reality will show you. You as an observer have eliminated the possibility of this paradigm and therefore won't experience it.

Like others studying and experimenting with consciousness, William Braud, a professor at the Institute of Transpersonal Psychology, wondered why some people could influence REG/RNG (Random Event Generator/Random Number Generator) machines while others couldn't. "He'd noticed several characteristics which tended to more readily guarantee success ... [and that] these effects work better if you believe they will and less than average if you believe they won't. In each case, like a REG machine, you are affecting the result—even if ... your effect is negative."[76] You restrict yourself to a particular reality, world, or way of thinking. However, if you allow yourself to be open to new possibilities, your reality will begin to reciprocate by providing you with evidence in your own experience.

Something I really think is essential in discerning what is right or wrong is illustrated in the motto of the IAC: *"Don't believe in anything ... Experiment! Have your own experience."*[77] I agree. *Don't* believe in something just because it's habitual or familiar. Be open to new possibilities, and use your own discernment as to what is true. This allows for all possibilities in your reality, and it makes you the central consciousness in your world—it gives you the authority to decide what is or isn't true for you.

In an energetic nutshell …

> *We are linked by a fabric of unseen connections. This fabric is constantly changing and evolving. This field is directly structured and influenced by our behavior and by our understanding.*[78]
> —David Bohm

Here it is in a nutshell: we live in and are part of an ocean of tiny, vibrating energy particles. These particles vibrate at an infinite number of frequencies through many different dimensions. However, for many of us, our experience of this is limited to the range of frequencies that our five physical senses can perceive. Not only can these minute subatomic particles exist in more than one place at a time, but they are completely interconnected "like some invisible web,"[79] have quantum holographic qualities that "[carry all] information everywhere,"[80] and only collapse into reality once our consciousness observes them.

Our consciousnesses have the opportunity and ability to choose our thoughts and, consequently, the beliefs that will ultimately result in the relative realities that we experience. This consciousness experiences life both through the physical body and through other, subtle bodies that are related to different dimensions and frequency ranges.

Because we (as individual consciousnesses) are at a rudimentary level of reality, we have all the more reason to take responsibility for what is immediately around us. If we can control what we choose to think and believe, which ultimately affects our reality, then our accountability is significant—both as a direct influence in our own lives and on a universal scale.

At the beginning of the book, I wrote, "Life is like a tuning fork, and the answer to everything is energy."So what did I mean by that?

That our consciousnesses are the tuning forks, and that each person's thoughts and emotions are the frequency of his or her tuning fork. What we resonate with is a direct result of the thoughts, beliefs, and emotions of our consciousnesses, or the frequency of our tuning forks. Resonance is a tool that helps us to measure reality. If we resonate with a reality that we are happy with, then we can be assured that our consciousness is collapsing the wave function to our advantage. If the reality isn't that great, then our tuning fork's frequency isn't serving us, and we need to adjust our thoughts and, ultimately, our beliefs. Don't worry if you're still a little lost, as I'll explain this more in the last three chapters of this book.

So now let's take this new paradigm and measure it against the questions I asked at the beginning of the book.

Chapter 3: Explaining the Unexplained

By now you should have a basic understanding of this new paradigm. It may be a little hazy at this stage, but hopefully you feel you comprehend the concept at an elementary level. Using this new outlook, I want to now answer those questions I raised in the first chapter of this book in order to further demonstrate this idea and solidify your understanding.

EXPLAINED CONNECTIONS

1. **How is it possible that twins, family, or even random people can be connected?**

A well-known study of a particular set of twins who were separated at birth astounded even nonbelievers. The following excerpt is from Louisa Gilder's *Entanglement*: "The pair of Ohio identical twins (both named 'Jim') separated at birth and then reunited at age forty... Their similarities were so striking that an institute for the study of twins was founded, appropriately enough, at the University of Minnesota in the Twin Cities. Both Jims were nail-biters who smoked the same brand of cigarettes and drove the same model and color of car. Their dogs were named 'Toy,' their ex-wives 'Linda,' and current wives 'Betty.' They were married on the same day. One Jim named his son James Alan; his twin named his son James Allen. They both liked carpentry—one made miniature picnic tables and the other miniature rocking chairs."[81]

So what causes this? Well, my belief is that it's a combination of entanglement and the holographic qualities of our universe. Essentially, the idea is that those tiny vibrating particles we've been talking about are all interconnected, and it's through this connection that holographic information is received in the form of signals. In the example above, when the two Jims made decisions throughout their lives, there was some subconscious plugging into the entangled, vibrating web of information that connected them together. Because twins are created together, this entanglement may be heightened; hence these experiences being noticed more with twins than other individuals.

Also, some twins may be more open to this connection simply because of their knowledge that they were created together so that they believe that, at a physical level, they should be more connected than other people. These twins are generally receptive to signals, feelings, or thoughts from the other twin. So we can imagine that if one twin were injured and in pain, his or her particles would suddenly vibrate intensely, and this sudden change in vibration would set off the interconnected particles of the other twin. The first twin might get a feeling or thought about the other twin and just know that something is wrong—almost like receiving a radio signal. Remember that this interconnectedness defies time and space and is therefore instantaneous.

There are some twins that don't experience this connection, and I believe it could simply be due to two things: a disbelief that it can happen and the fact that their senses aren't tuned into this entangled web. Both of these facts aren't absolute and have the ability to be changed if they are willing.

Interconnected experiences like these should be similar for any related individuals since we're all made of the same stuff, so to speak. But what about when a friend you haven't seen in years—maybe even just an acquaintance—suddenly pops into your mind and you then get a call from the person or bump into him or her? This results from the same condition I've been describing: everything is interconnected. As mentioned earlier, anything made up of atoms (which is everyone, everything, every space in the universe), is made up of these tiny, vibrating strands of energy. You are therefore connected to all the millions of people you have never met, but because you don't really having any recognition of this on a conscious level, the signals may never be registered or translated—or if they are, you might not realize it. "The more intense the emotion, the greater the signal intensity."[82] If you have a close relationship, the signal is more intense and familiar than that of someone you're not as close to.

In fact, it has been proven that some form of interconnectedness does take place regardless of whether you know someone well or not. In an experiment performed by Dr. Charles Tart, a researcher in the field of consciousness who holds a doctorate in psychology, Tart administered electrical shocks to himself and then attempted to telepathically transmit the pain to a person in another room who was hooked up to machines that measured physiological changes. Although the recipient wasn't consciously aware of the pain, it registered in the person's physiology.[83] Virtual strangers were interconnected purely by intent. The vibrational signals produced changes in the recipients' physiology, the probable difference being that, had they known each other, they might have translated the same vibrational influence into a feeling or thought about the other person.

I guess you may be wondering what is actually happening when you think of other people and then bump into them. Neither of you consciously decided to think about each other—so why did that pop into your heads? Well, your brain energy particles don't register your frequency because of any specific thought, but on a deeper level, you recognize and resonate with each other's frequencies. If you are open to reading these vibrations and translating them into thought, then there's a good chance, if you're in the vicinity of the other person, that you might think of him or her and bump into the person a few minutes later. Or if your friend or acquaintance happens to be in the next aisle in the shopping center, you may just pick up on that specific frequency as a result of entanglement, causing you to think, "I really should give that person a call." Later, you might be talking to each other and realize that you were both thinking of each other and had both been shopping at the same time and in the same place. Coincidence no more—it's now been explained.

2. **Why do some people experience this synchronicity more than others?**

The answer to this is that it simply comes down to whether you are open to believing it or not and whether you are sensitively aware of the connections. If you remember, I mentioned there was a catch to this paradigm. You have to believe it to see it. In the experiment by William Braud, participants were more successful at influencing the REG/RNG results if they believed they could. Even those that didn't believe affected the outcome in a negative way, so their results still proved the theory that their minds had influenced the REG/RNG machines by mirroring their belief pattern, which in this case was nonbelief.

What does this mean for those who don't believe in these things? Simply that their beliefs will affect their reality. In the example of the REG/RNG experiment, the individual's nonbelief shut down the receptors that would translate vibration into feelings or thoughts. A closed mind closes down the possibility of experiencing entanglement as a macroscopic reality. We all have the ability to access this web of entangled information, but we may have been taught to ignore our intuition.

If you remain open to the possibility, you will allow yourself to start translating these vibrations into your experience. This might take a while, especially if you were originally closed to these ideas. The vibrations and signals will be there, but because your ability to translate them has been turned off, it might take a while to learn to respond to what your intuitions (internal vibrations) are telling you. It's like learning a whole new language, except that this one is more difficult to teach because people receive vibrations in different ways. Some people get visions, which are internal vibrations that have been visually translated by our minds; others might hear a name, register a feeling, or just have a thought pop into their heads. The key is to start being aware of what is going on around you and internally.

If you are a believer but feel you still don't sense these intuitions, there are exercises you can do to increase the range of frequencies your senses pick up. (I talk more about these exercises when answering question 4 in the part headed Normal Activity) It is also all about becoming aware of what is going on around you. If you have ever walked into a room after an argument and could feel the tension, realize this is a form of your senses reading the field of information and sensing the change in the energy vibrations. Almost everyone senses some vibrating energy particles; many just aren't aware of it. Now is the time to open your mind, awareness, and sensitivity to it all.

3. How can connecting with one person make you feel drained while connecting with another energizes you?

There are two types of relationships: those that resonate with you and those that annoy you to no end. Anyone that you consider close to you, such as a good friend, partner, or family member, generally resonates with you on some level or in some specific way. For example, you may have sports friends. You generally only hang around them to play sports, talk sports, or socialize around sports—such as going to a game or sports bar. Then there are those friends that resonate with you on a purely social level. They know how to party and you always have fun with them, so when you are resonating on that frequency, they are the people you call. Still others you call when you need to discuss your goals, move forward with your life, or talk on a deeper level. Again, you probably have specific friends that you go to for this level of resonance—but perhaps not to play sports with.

Then there are those people that resonate with you on more than just a few levels. They resonate with you in all aspects of your life, including your goals, values, and beliefs. These are generally your partners or closest friends. You feel that they are on the same wavelength as you—much more so than other people. You feel that you can read each other's minds sometimes and that each knows what the other is thinking. These people are heavily interconnected with you. Generally, when you are around these people, and especially when you talk about or are doing something that really connects and resonates with both of you, you walk away buzzing. You were truly harmonizing with each other and raising each other's frequencies.

What's happening here is simply that your personal frequency matches the thoughts and feelings of the other person. You tune to each other and make amazing music—in the same way that an orchestra, when it plays a piece of music perfectly, can raise the energy levels in the room. When musicians are out of tune, miss on the timing, or even play the wrong section of the music, it sounds harsh, distorted, and unpleasant to the ear—not something you'd want to participate in. By analogy, our minds, bodies, and souls are the instruments, and when we find other instruments to harmonize and make beautiful music with us, the feeling we get can be ecstatic. So when we walk away from a close friend, partner, or anyone that we resonate with strongly, we generally walk away with a raised frequency—we feel elated. Next time this happens to you, stop and really feel the presence of this vibrating energy. Once you really become aware of it and tune into it, it will be easier to come back to that frequency over and over again.

On the other side of the relationship coin, there are those people that irritate you to no end—they completely annoy you or really peeve you off. Again, this has to do with what is going on at the deeper Planck level. The frequencies of your tiny vibrating energy particles may not be matching each other on any obvious level; instead, they are resonating with you at a negative frequency. What this means is that they are resonating with a part of you that you are not happy with, and that's what causes the negative feelings. It is a reflection of reality at its most blunt. It's these people that you have to watch out for, because they are showing you a frequency of yours that bothers you. They come into your life to illustrate a point. So when you come across these people that conflict with you, be grateful, as they are showing you an area that you need to work on. That is why it irritates you so much: it's another form of resonance but at a deeper frequency level that you've hidden from yourself. Penney Peirce puts it well: "It's great to realize you're similar to people you admire, but you may not want to admit that you could be anything like the people who irritate or threaten you."[84]

I'll give you an example that happened to me. I knew deep down that I was a bit of a bossy person—that I could be a bit controlling, snap at people if things didn't go my way, and be very protective of my ego. In fact, if someone seemed to be illustrating a flaw in something I had done, rather than embracing this new knowledge and learning from it, I would defend it to the ground—even when I was making no sense! Along came Sandra,* who was exactly the same kind of person. Well, funnily enough, I really didn't get along with her. She annoyed me, and we locked horns on so many things it was ridiculous. My instinct was to run away, avoid her, and just continue as I had. Then one day a friend pointed out that we were actually quite similar. I must have been ready to hear it, because rather than denying it and arguing until I was blue in the face, I listened and started laughing at how blind I had been.

I decided from that day on to make a concerted effort to find all the things I liked about Sandra. At first it was impossible. It actually put me in a bit of turmoil. My insides were saying, "How can you try to like this person—I am busy trying to point out that you are nothing like this person by hating her." I knew that it was just me trying to protect myself, but in fact it was holding me back and preventing me from growing. Every now and then, I would fall back into an argument with Sandra, but over time I realized that I could actually like her and even get along with her. We could even have a laugh. It was amazing. The more I focused on the things I liked about this person, the more she seemed to display likeable qualities that I never would have noticed if I had still been focusing on the things I disliked. It is funny that now even when we have small clashes every now and then, we can still have a laugh about it afterward and learn from it. Everyday the relationship gets better. Now this person no longer irritates me—I just accept her for who she is. Our frequencies may not resonate on a strongly positive level, but what is exciting is that they no longer resonate on a negative level—which shows me that I have removed that frequency from myself.

So what about those people who drain you? This is a slightly different kettle of fish, and it's not necessarily about resonating frequency. It has more to do with energetic connections—another level in the complex system that makes up who we are.

We connect with every person we talk to or hang out with during the day. There is generally an exchange of thoughts, ideas, and even emotions.

---

* Not her real name.

On an energetic level, these things are quite palpable, and we can walk away feeling heavy when we've had someone else's stuff dumped on us! What may have happened is that they connected with us, and because we allowed it, they downloaded their negative energies into our energy body. They literally sent us palpable negative energies, which they then left with us. The result was that, though they may have felt much better and lighter, we walked away feeling heavy and negative because we allowed these energies to affect our own. There is a way of preventing this from happening, which I will talk about in chapter 4.

Another form of negative energy exchange has to do with energy vampires. These are the people that go around, mostly subconsciously (these people are usually completely unaware that they are doing this), draining us of our own positive energy. We may be feeling completely fine and happy and looking forward to seeing these people, but once we start conversing with them, we feel like all our positivity is being sucked out of us. It doesn't matter what you say or do to try to make this person feel positive or happy. It's not solutions they're looking for—it's your energy. They do this to many people around them, because they are unable to fill themselves with or access this energy in any other way—either because of their upbringing or purely from not knowing how to do it.

I have been on both sides of this. In my teens, I remember being a bit of an energy vampire. I would have conversations with my friends in an effort to cheer myself up, but I was basically fishing for compliments. I would try to get them to say positive things about me because I couldn't see them myself, but often I would verbally flick back any compliments they gave me in order to get more. Subconsciously, I wanted them to shower me with the positive energy from their compliments in order to try to build myself up. But it didn't matter what they said. Once they were gone, the energy would deplete, and I would have to find someone else to help me feel stronger and more positive. I ended up, over many years, realizing that there is only one source of endless energy: one's self.

On the other hand, I have had friends that have drained me more or less in the same way. At first I fell into the same circular trap, feeding them with advice and compliments—anything to try to help them. Every time I did this, I realized I was being drained and had to build myself up again. I continued helping them, as I wanted to see them happy and positive. The problem was that I kept getting sucked back into the metaphorical hole with them so that I was then unable help them. I ended up being brought down to their frequency level. They sucked away all my energy so that

I became low, and because, as a result of their insecurity, they couldn't maintain their own energy, they remained in that condition. I had to start helping them out of this hole by not getting into it with them in the first place. I offered them advice designed to provide them with a ladder or rope to help them get out of their hole, but they ignored it. They wanted my energy, not my advice. So after a while, I found myself having to make a decision: if these friends weren't going to attempt any of the advice I gave them to get themselves out of this hole, and if they were going to continue to drain me, I would have to let them go. As horrible as that sounds, it was the only way to stimulate them to try to find another solution, as their energy source was now gone. It also prevented me from being caught up in it all and, as a result, being unable to move forward myself.

If you are confident and believe in yourself, then look proudly at yourself in the mirror and tell your reflection how much you love it. There is nothing more energizing than that! And it won't matter how many times someone tries to drain your energy—your supply comes from within, which means that you radiate energy infinitely. In chapter 4, I'll talk more about how to reach this state as well as how to stop people from draining you of your energy.

4. **What is going on, other than biology and chemistry, when there is a love connection?**

When love occurs, what is going on at the deeper Planck level of our existence is, again, resonance, an intense form of frequency matching. As Penney Peirce says, "Those who populate your world *frequency-match* with you."[85] As I explained in the previous chapter, we have many tiny tuning-fork frequencies that make up our own complex individual frequency. When these combined frequencies match with another person on every level, then a unique form of resonance takes place that raises both people to a higher level. Again, Peirce: "When souls have natural resonance, they simply appear to each other."[86]

So although there may be animal instincts involved, at the deepest level, we have tuned into a person that resonates with us on many levels. This is what is going on in a love relationship that sets off chemicals that give us a high. When we feel really connected with someone, it is literally because they are on our wavelength!

So what happens when we fall out of love?

As I've said before, people are vibrating energy particles with many different frequencies occurring all at once. Like our minds, which can have ten different thoughts a second, our frequencies can change. During a single day, we can feel high, then low, upset, or angry, and then high again. Over time, our frequencies evolve as we evolve through our life experiences. You may fully resonate with someone at one time so that you feel that you are connected, that you are soul mates, and no one could ever convince you otherwise. Five years down the road, if you don't align your goals and values at the start, you may start drifting apart, both mentally and on the level of frequency. When one or more of your frequencies changes because of new life experiences and goals, something both of you felt connected with—such as the idea of having children—may no longer be as important to you. Your thoughts and emotions shift, so your energy frequency, and eventually your reality, shift along with them. Your partner no longer resonates with you—you have become out of sync. Even though you both have that history and connection of love with each other, the foundation of your relationship—your mutual resonating frequency—is no longer structurally stable. A few of the frequencies no longer harmonize, and the song that was your relationship has started to sound out of tune and off-beat.

Resonance is the key to love. Yes, there are definitely biological and chemical influences as well, but even these are made of vibrating energy particles—which brings us back to resonance as the key. It is the foundation of a relationship, so to remain on a resonating frequency, you need to constantly communicate and work on your frequencies together. Again, I'll talk more about this in chapter 4.

To find this soul mate, you need to be present in your own true frequency. Be who you truly want to be. When you resonate at your truest frequency—your tuning-fork frequency—you will naturally attract others on the same wavelength, like being tuned to the same radio station. It's with these people that you will form strong bonds. As they say, "Birds of a feather flock together."

5. **Why do we connect with some activities and not with others?**

What we resonate with at any given time depends on our frequency levels, and these can change often as a result of the thoughts we think. Think back to chapter 2, where I talked about thoughts and emotions affecting the energies that change our reality. Just as we can think a hundred thoughts in a minute, our frequencies can change quite regularly.

Let me give you an example of an experience I had that demonstrates this principle. I have always been a reasonably good basketball player. I am no Lauren Jackson, but I've achieved a few basketball awards and been chosen for quite a few elite squads, so I was really confident. When I went to university, I decided to continue with this passion of mine. I was still in my generally insecure phase at the time, and the girls that played on my team started to feel threatened by my skills or maybe unhappy that their team was changing. They focused on my mistakes, pointing them out—so much so that I, too, started to focus on them. Eventually, it got to the point where I had become so insanely crippled by my thoughts that my reality mirrored my mental state and I couldn't even catch a ball. It got beyond ridiculous. I stopped playing basketball and didn't go back for six years. Luckily, when I started playing again, my brain had rewired itself back to my natural skills, and because I had become a confident person, I was able to revert to the level of playing basketball that I'd once known. This shows you that belief in your own skills and abilities can change through new habits of thought—so why not change your thought habits to positive rather than debilitating ones?

You can see from this example that even though I resonated with the overall activity of playing basketball, that same activity could be affected in different ways by my thoughts. How I interacted with it—whether it worked for me or not—always depended on how I thought and felt about it, and this goes for any activity, job, or career.

You may also find that you at first completely resonate with a job, such as recruitment or driving a forklift. Everything about it resonates with you—from learning it as a new skill to the things that you ultimately accomplish. Again, this has to do with the thoughts and feelings that surround the activity. It's these that resonate with you, creating the connection that you feel. For example, there are times when we don't connect because of experiences we've had as children. Maybe you have always wanted to learn how to dance, but as a child, you were told you were terrible at dancing and that your dancing would never amount to anything. As a result, your dance frequencies have been prevented from fully resonating. I guess it all comes down to your interpretation of an activity. Some people get a buzz out of cleaning toilets. They feel a sense of accomplishment—that when they walk away, the place they've just cleaned is as clean as it can be. They completely dissociate themselves from what are often viewed as the negative aspects of the task, and this allows them to resonate with the job of cleaning toilets. In the same way, some people absolutely love statistics, counting, formulas, and numbers, which to others seem like a complete nightmare. Again, it has to do with our interpretations—of reality or of our experience of it. It has to do with our thoughts and our feelings, which come from our consciousness. So if you want to resonate with something that you don't like naturally, you have to change your thoughts about that thing—you have to change your concept of that reality. If you really think, "You know what? I want to give statistics a go," you'll have to change your thoughts to match that frequency, and ultimately you will find yourself resonating with it.

COMMON ANIMAL BEHAVIOR

1. How do pets know the intentions of their human owners?

Rupert Sheldrake did a lot of research into the otherwise unexplained powers of animals, especially in the areas of intention, connection, precognition, and the animals' ability to sniff out ailments. His main conclusion about how animals can do a lot of this has to do with the concept of morphic fields: "A wide range of unexplained powers of animals might be explicable in terms of morphic fields."[87] In other words, pets are able to tune into and pick up on the intentions of their owners through the entangled holographic field, which provides them with information about their owners' thoughts and feelings.

Animals are interconnected with us just as we're interconnected with each other. This phenomenon is very similar to the one mentioned in the first question in the first part of this chapter, which describes the interconnectedness between twins. Because of the closeness between pets and their owners—which sometimes even resembles a human relationship—experiences of entanglement with animals can be quite strong. As soon as a human being makes definite moves toward coming home, for example, the animal translates the signals by picking up on the owner's intentions. I believe that animals are more open to translating these signals because they have no interfering thoughts or things to distract them from interpreting them in any other way. In other words, they receive the signals instinctively.

Have you ever thought, for example, "I think I'll walk the dog," and then seen the dog get all excited? It knows when your intention is sincere. It is picking up on the energy of your intention through morphic resonance.

Dr. Edgar Mitchell also feels that morphic resonance—this ability to pick up on or tune into individual morphic fields—may very well explain this amazing connection between human beings and their animals. Mitchell tells another interesting animal story that relates how his father, a rancher, could pick up when something was wrong with his herd. "He could intuitively recognize if one of his herd was in trouble. In the middle of the night he might awaken and go searching for a cow that was having difficulty delivering a calf. And invariably she was. He would unerringly find her in the middle of the night even though she had hidden herself in the brush for safety. These nonlocal resonances that are the most basic of nature's information management schemes work not only within a species, as biologist Rupert Sheldrake proposed, but between species as well."[88]

One of the most intriguing features of these morphic fields is that members of different animal species show that they learned from the experiences of other animals within the same species by downloading the information through the process of morphic resonance, which transmits the information through the morphic field. What do I mean by that? A famous example is of the blue tit bird. Members of the species in one area learned that they could tear off the lid of a milk bottle and drink the milk. Soon after the birds had learned do this in one area, their counterparts in regions too far for the blue tit to fly began to adopt the same behavior. They were picking up this new habit from the morphic field via morphic resonance. Other animals illustrated the same behaviors. "Several ranchers have told me that herds not previously exposed to real cattle grids will avoid the phony ones ... Sheep and horses likewise show an aversion to crossing painted grids. This aversion may well depend on morphic resonance from previous members of the species that have learned to avoid cattle grids the hard way."[89]

These morphic fields are providing a network of information that animals can detect through morphic resonance. It is almost like an intuitive form of the internet, except that rather than having to Google for information, the animals pick it up purely by instinct. They also have the advantage that they aren't affected by external influences about what they should believe. They just perceive the information and react instinctively.

2. **How do animals sense the medical ailments of humans?**

Science has shown that animals can register a much vaster range of frequencies than humans. We all know that dolphins have a high-frequency range and that they use the same frequency as a radar to get around under water, to know when danger is near, and to find food. Dogs can hear really high frequencies, which explains those dog whistles that you can buy that are beyond the normal human range of frequencies but are annoying to dogs.

This same idea applies to the ability of dogs to pick up on medical ailments in humans. The theory is that they do this through the morphogenetic field, which enables them to first sense the unnatural frequency (the ailment) within the human, which tells them that the human's blueprint isn't quite right. "These [morphogenetic fields] are rather like invisible blueprints that underlie the form of the growing organism ... They are fields: self-organizing regions of influence, analogous to magnetic fields and other recognized fields of nature."[90] They then use their larger range of frequency to sniff out or intuitively know exactly where the problem spot is located on the human being's body. Rupert Sheldrake illustrates an example of this from his research of specific incidents:

> The patient first became aware of the [cancer] lesion because her dog (a cross between a Border Collie and a Doberman) would constantly sniff at it. The dog (a bitch) showed no interest in other moles on the patient's body but frequently spent several minutes a day sniffing intently at the lesion, even through the patient's trousers. As a consequence, the patient became increasingly suspicious. This ritual continued for several months and culminated in the dog trying to bite off the lesion when the patient wore shorts. This prompted the patient to seek further medical advice. This dog may have saved her owner's life by prompting her to seek treatment while the lesion was still at a thin and curable stage.[91]

At the beginning of the book, I mentioned a special clinic in Atlanta where disabled children go swimming with dolphins that can sense whether or not and how much these children are able to play with someone. Playing with the dolphins results in improvements in the children's condition. Again, I believe the dolphins at this clinic pick up on the frequency of the human beings. They act just like a clear signal from a radio receptor—no interference, no misinterpretation. Whatever information they receive, they act on it accordingly, whereas humans might receive the information but then misinterpret it because of past experiences, beliefs we've learned, or just pure misjudgment.

I believe that the disabled children benefit from some form of resonance. The experience with the dolphins raises their frequency levels so that, by resonating with their environment—the dolphins—they are able to experience a pure, natural, healthy frequency, which helps to increase the overall health of the children. There is no frequency of sadness coming from the dolphin toward the child—none of the angst or negative emotions that can naturally occur with parents because of the struggles they go through with their children. The dolphins provide a nonjudgmental, high-energy frequency that the children can resonate with and that therefore helps to increase their wellness.

This kind of thing also happens with pets around the house. I have heard of studies saying that families with pets are less stressed. Like the dolphins with the disabled children, pets can provide a pure source for a healthy, loving energy frequency that resonates throughout the household. Yes, there are psychological benefits too, but I believe that the foundation for this stress relief comes from entanglement of frequencies through the morphogenetic field of pets and their families.

3. How do animals sense paranormal phenomena, such as ghosts?

This comes down to two things: the larger range of frequencies that animals can perceive and the multidimensional nature of our universe. I will go into more detail in the next section about paranormal activity; however, to give you an idea, I want you to think back to chapter 2. I've talked about the multidimensional nature of our universe, the fact that right now there are many different frequencies, and therefore dimensions, taking place in the same room. We can't pick these up with the limited range of frequencies we have through our five senses. Animals, as mentioned before, have a wider range of frequencies that they can pick up and can therefore also pick up on other dimensions that many human beings can't perceive. This is essentially what I believe is happening when animals sense paranormal phenomena: they are perceiving the vibrations of higher frequencies.

An experiment done in 1977 by an American researcher, Robert Lyle Morris, was designed to see whether an animal could perceive someone during an out-of-body experience, implying that they could sense presences from other dimensions. In his book *Projectiology: A Panorama of Experiences of the Consciousness Outside the Human Body,* Waldo Vieira writes. "the results suggested a *parapsi interaction* between the animal and the projected consciousness of the student."[92] I will talk more about these subjects when I answer the questions on paranormal activity; however, from this I'm sure you can see that an animal, in this case a two-year-old kitten, detected the subtle body of its owner when projected outside of the body. Using its larger range of perception, the kitten was able to see into the subtler dimension.

4. How is it possible that animals can predict earthquakes, tsunamis, and even bombings?

Rupert Sheldrake came to two possible conclusions when talking about animal precognition: "As far as I can see, only two possibilities remain: 1. Telepathy. The animals picked up influences telepathically from people or animals along the flight path of the bombers ... [or they] might have picked up the hostile intentions of the German bomber crews ... 2. Precognitive forebodings. Perhaps the animals somehow intuited what was going to happen in the near future."[93]

Essentially, the animals used their natural instincts to pick up the disturbances in the morphic field or via entanglement with holographic information and were able to perceive the signals as a sign of danger. Translating these signals into an action seems to be instinctual for most animals.

Once we develop an understanding of morphic fields and morphic resonance, of particle behaviors and entanglement, and of the holographic nature of the universe, it all starts to make sense. For example, the elephants that ran to high ground an hour before the tsunami arrived may have been subconsciously interconnected across the morphic field to the particles in the ocean or even to fish that may have been signaling a disturbance. I'm not sure exactly how the elephants managed to interpret the signals—it could have been pure instinct. Their own vibrating particles must have intuitively told them to run in a certain direction or to get to higher ground, and that may be why they fled. To sum it up, according to Sheldrake, "the morphic field hypothesis may be able to account for a wide range of unexplained powers of animals, both telepathic and directional."[94]

This may all seem quite out of this world right now, but many highly trained scientists have embarked on various investigations that show us many things that are beyond the current perception of reality held by many people. It's up to us to be open to the new and exciting changes that are upon us. Even though these ideas may seem ludicrous at first to some, it is hard to deny that the answers may very well be staring us in the face, especially when everything starts to make sense. We just need to be willing to fully open our eyes to new perceptions.

NORMAL ACTIVITY

1. **Could ghosts be real, and if so, how are they possible?**

The simple answer to this is yes, because I've felt it to be so. However, this doesn't satisfy the scientist in me. It doesn't help me to understand how that could even be possible. So let me break it down for you—for my own sanity and so that you can understand:

The fabric of the universe is an ocean of vibrating energy. The frequency of this vibrating energy depends on the dimension in which we find it. Most of us are limited to seeing only our own dimension or reality, because our five senses are limited to a certain range of frequencies. We are consciousnesses. It is one's consciousness—soul, spirit, or true essence—that manifests through one's physical body alongside a few other, subtler-frequency bodies.

The research I mentioned under the subheading Consciousness in chapter 2 has found that there are three other bodies. The *holochakra*, which is the energetic body, includes the aura and chakras, which I will talk more about in the next question. Next, there is the psychosoma, the emotional or etheric body, which is where consciousness lies as an extraphysical consciousness (outside of the physical body). This is explained further here:

> Surrounding and interpenetrating our physical body is an energy field or structure known as the "etheric body." The etheric body is a kind of invisible duplicate of the physical body that actually occupies the same space as the physical body (but at a higher vibratory rate or energy frequency than the physical). It is the first in a series of what are called the "higher spiritual bodies." In a very real sense, our soul, our "true self," expresses itself through a physical body that is subtly influenced and molded by these various higher spiritual bodies. Each of our spiritual bodies is formed from vibrating life-energy fields of progressively higher and finer levels of energy and matter.[95]

Finally, there is an even higher subtle body called the *mentalsoma* or mind. It is in this body or dimension that people experience samadhi, nirvana, or some form of enlightenment. It is so subtle and at such a high frequency that it's only when the consciousness is in this body that these states are ever attained. Physical bodies and even etheric bodies are too dense for this extremely subtle dimension.

"In the vibrational universe it is believed other dimensions exist, beyond the physical plane, that our spirit normally inhabits during its natural state of existence. In fact, we all explore this higher realm when we sleep at night. While our body sleeps and our mind enters the dream state, a higher aspect of our spirit leaves our body and has direct experience of this higher-dimensional plane of existence."[96] This is known as either lucid dreaming or, in the case of a total experience, an out-of-body experience, which I will talk about later.

When we die, we shed our physical bodies, but as our consciousness is something other than matter/energy, it survives this physical death and maintains its presence in the emotional body. Richard Gerber says in *Vibrational Medicine for the 21st Century* that "in the vibrational or energetic model, spirit, like a vaporous ghost, inhabits the mechanical vehicle we call the physical body. At the time of death, our spirit moves on, leaving behind only a lifeless shell."[97]

If we are aware we have died, we move onto subtler dimensions where we prepare to return to this dimension in order to continue the evolution of our souls through reincarnation. Some children and even some adults have remembered their past lives on being reborn. The research shows that there are quite a few people who do, giving further evidence of this paradigm. Amit Goswami notes that "University of Virginia psychiatrist Ian Stevenson has accumulated a database of some 2000 such claimed-reincarnational memories which have many characteristics that have been verified."[98] He goes on:

> Consider the case, studied by L. Hearn at the end of the nineteenth century, of the Japanese boy named Katsugoro who at age eight claimed to be Tozo, the son of a farmer in another village, in another life a few years earlier. He also said that his father had died when he was five in his previous life and he himself had died a year later of small pox. He gave many details of his previous birth, for example, the description of his parents and of the house in which they lived. When Katsugoro was taken to the village of his previous life, he was able, unaided, to find the house in which he lived then. All together, sixteen items of his past-life recall checked out.[99]

There are many of these stories, and you only have to search libraries, the internet, or even YouTube to watch amazing documentaries of children undeniably remembering their past lives.

Goswami further writes that "Edgar Cayce ... [was] able to read other people's reincarnational past lives ... Cayce, under hypnotic sleep, gave about 2500 such past-life readings, sometimes more than once but never contradicting himself."[100] To explain how this could even be possible, Goswami expresses his belief that, through entanglement, we are able to access holographic information. "How could a person like Cayce look into the nonlocal window of another individual? Cayce's own answer was 'Akashic memory,' for which an acceptable translation is nonlocal memory, but I think a more tangible explanation exists in terms of the nonlocal window of our model. The point is that, in principle, consciousness is one; thus, any person's nonlocal window that connects all of her incarnations is open to everyone who knows how to look, but this is a very unusual capacity."[101]

Many people are unaware at first that they have died. This might be because of our belief that when we die, we don't exist—therefore, if we still exist, we must not be dead. This is like Bruce Willis's character in *The Sixth Sense*. Author and researcher Marie D. Jones elaborates on this in her book *Modern Science and the Paranormal (Haunted: Ghosts and the Paranormal)*: "Ghosts remain in this in-between land because ... they have unfinished business. Perhaps the trauma or emotional turmoil involving their death creates a shock to the system that does not allow them to 'cross over,' as a popular medium named John Edwards likes to call it."[102]

Robert Monroe, author of the book *Journeys Out of the Body* and founder of the Monroe Institute, began scientifically analyzing the perceptions he derived through his many OBE experiences. He came to the conclusion that there existed a dimension—the dimension closest to our own—which he calls Locale II. According to Monroe, the "scope of Locale II seems limitless. Under the conditions encountered thus far, there seem to be no means to measure or calculate the breadth and depth of this strange familiar place. Movement from section to section is too instantaneous to allow any estimates or to observe relative spatial positions of one area to another."[103] Waldo Vieira, the founder of the IAC, calls Locale II the "*paratropospheric dimension*—the extraphysical sphere that coexists with human life—[that] constitutes a fluid, plastic, nonphysical and apparently omnipresent environment. Although formless per se, this dimension has the property of taking on or reflecting whatever form is mentally impressed upon it."[104]

As Monroe notes, this dimension, or Locale II, seems to be closely connected to our dimension, and there is evidence of frequent crossover between the two. "Joshua P. Warren, … president of LEMUR (League of Energy Materialization and Unexplained Phenomena Research) … told [Marie D. Jones] that just as some aspects of humans exist in another dimension (one of mind), ghosts seem to primarily occupy some other dimension as well … 'In fact,' he stated, 'they may primarily reside in the same dimension as that of our minds. Understanding how and why they manifest relies on both understanding that realm and determining what conditions in our physical realm bring those two dimensions temporarily closer.'"[105]

What brings them closer together is the ability to perceive the different frequencies. Animals or people with the ability to perceive a higher range of frequencies—such as clairvoyants or clairaudients—have been given a very negative stigma, when in fact they have an unusually large range of senses. It is interesting that individuals who possess abilities out of the ordinary are either admired, like top athletes, or disbelieved, like clairvoyants, clairaudients, or other types of psychic. The only difference is that one is achieving something within our realm of understanding, while the other is reaching beyond our normal paradigms.

The good news is that science has actually started catching up and has even photographed the ghost, or etheric body, of plants. Barbara Ann Brennan writes in *Hands of Light*, "Through the methods of Kirilian photography, people have been able to photograph an entire leaf after half of it was cut away."[106] Gerber adds, "The 'phantom' in the phantom-leaf photo looks identical to the missing part of the leaf that had been destroyed … down to the inner vein structure and outer leaf geometry."[107] This is photographic evidence of the etheric body of the plant. It would be interesting if science took this further and was able to photograph a human limb that had recently been amputated. Although not really ethically or currently possible, it could show us the etheric body of a human—essentially, a human's ghost body.

So ghosts, or extraphysical consciousnesses, do exist. As Warren explained to Jones, "Ghosts are 'non-physical entities' if we define non-physical as 'not being restricted to the known laws of physical matter.'"[108] They exist because we exist—they were once us. That doesn't mean that you now have to freak out because I am saying that there are ghosts—there always have been. If you're not a person that is open to this, you probably will never experience anything. You may have restricted your ability to perceive that range of frequencies. If you do become open to this, don't stress about it, as it is much like when we resonate with people of similar frequencies in our normal dimension—the same goes for the extraphysical dimension. Only if you are of a dark, negative nature will extraphysical consciousnesses of the same frequency tune into your frequency. If you want to change that, raise your frequency and think of all the positive things that make you happy. Only then will positive extraphysical consciousnesses resonate with you.

2. **What are angels, and how is it possible that they even exist?**

Although you may think I'm starting to sound a bit airy fairy again—and for those of you who may be completely lost by now because I've started talking about angels—really try to get rid of those negative connotations that you may have with the word *angel*. Try to open your mind. You can even try thinking, "Okay, have I ever experienced something that I cannot explain?" and see whether this paradigm can actually explain the experience.

Refer back to the paradigm in the previous question about ghosts for a moment. I want you to think about the in-between phase of our lives—what's known as the transition between life and death. I want to combine the multidimensional aspect of our reality, the idea of extraphysical consciousness, and something that has to do with energy vampires, which I talked about in the section on relationships. Mix them together, and you will have some answers—and maybe more questions!

Okay, but seriously, here is what I've learned:

As we have more and more lives, we are supposed to be evolving as consciousnesses. There are extraphysical consciousnesses that are between lives that are slightly more evolved than others. They are there to help us during our intraphysical life. Helpers (as they are referred to by the IAC), or angels, are there as guides to support us through our evolution. In no way do they influence us—it is purely our own free will that decides what we do in life. However, if we are on the right path, they guide us by perhaps making us notice something that will help us on our journey. Whether we take that information on board is entirely up to us.

When my friend was prevented from being run over, the angel may have done so because the alternative wasn't in my friend's soul purpose. We may wonder, in that case, why bad things still happen to us. Well, if we stray from our path and go in a direction that doesn't help us, then the helpers will obviously leave and help other, more committed souls. They don't waste their time with people who aren't moving forward. Also, if you are constantly sick or injured, perhaps your thoughts, beliefs, and realities aren't supporting you on your evolutionary journey.

Amit Goswami describes these helper beings in other terms: "Those karmically fulfilled quantum monads that are 'reborn' in Sambhogakay form are what we call angels. They are available to help others through channeling, although they may never return to incarnate in a physical body."[109] Essentially, we all have our theories and experiences with angels; however, these will be directly influenced by our beliefs and our sense of reality. So although there is evidence of extraphysical consciousnesses, your experience of them may be slightly different from mine because of our religious, cultural, and other beliefs. It is all in the eye of the beholder. Overall, though, the evidence through research, scientific experimentation, and personal experience is pointing toward a world with extraphysical consciousnesses—some that help you, such as angels, and others that are on their own path, unaware that they have even passed on.

3. **How are psychic abilities possible? How can the future be read if it hasn't happened yet?**

Psychic abilities, such as precognition (seeing into the future) and remote viewing (seeing things at great distances beyond the usual human range), have been scientifically studied for quite some time and even have their own field of study. Marie D. Jones describes this trend in her book *Modern Science and the Paranormal:* "People of every background and culture believe in the mind's power to reach beyond the five senses and receive information from sources outside the boundaries of space and time. Also referred to as 'psi' (the collective term for paranormal and psychic phenomena) in the word's truest sense, psychic phenomena have become so widespread and pervasive that serious researchers, including many open-minded scientists, have created a field of study called 'parapsychology.'"[110]

Current research is leading to the recognition that psychic ability is a part of everyone's makeup. It's only whether you are open to it and how much you have used it that determines its place in your life. In his book *The Holographic Universe,* Michael Talbot writes, "In a now famous series of experiments physicists Harold Puthoff and Russell Targ of the Stanford Research Institute in California found that just about everyone they tested had a capacity of 'remote viewing,' the ability to describe accurately what a distant test subject is seeing ... Puthoff and Targ's findings have been duplicated by dozens of laboratories around the world, indicating that remote viewing is probably a widespread latent ability in all of us."[111] Even the CIA was interested, since they knew that the Russians were already putting a lot of money into it, so they asked Puthoff and Ingo Swann (a psychic) to prove themselves. Despite the fact that the CIA sent them out into the field, it later denied their success, probably out of a fear of ridicule so prevalent at that time, even though other countries were also involved with these experiments.[112]

It isn't just recently that evidence has been compiled, either. Abraham Lincoln is reported to have predicted his own death: "Abraham Lincoln had predicted his own demise only three days before his actual assassination at the hands of John Wilkes Booth. Lincoln reported the dream to author Ward Hill Lamon, who recounted it in his book *Recollections of Abraham Lincoln 1847–1865*. Lincoln's dream contained enough specifics that it is chilling to read it in light of what was to happen."[113]

How is all this possible? How can people predict the future or see things at a distance? Marie D. Jones believes in two possibilities, and I tend to agree with her second theory: "The basic theory behind RV [remote viewing] suggests that information is somehow transmitted from sender to receiver via the electromagnetic spectrum. Another theory suggests that remote viewers are accessing a sort of universal matrix of information that is available to anyone with the skill to tap into it."[114] Dean Radin, senior scientist at the Institute of Noetic Sciences, has also come to believe the latter theory after compiling a few theories, which he sums up as follows: "field theories include ... Carl Jung's idea of the collective unconscious, biologist Rupert Sheldrake's morphogenetic fields, and neuroscientist Michael Peringer's geomagnetic field theory. These models all postulate the existence of some form of nonlocal memory permeating time and space that we can resonate with."[115] Finally, as reported by Michael Talbot, "Both Puthoff and Targ feel that nonlocal quantum interconnectedness plays a role in precognition, and Targ has asserted that during a remote-viewing experience the mind appears to be able to access some kind of 'holographic soup,' or domain, in which all points are infinitely interconnected not only in space, but time as well."[116]

To sum it up in my own words, anyone with the tendency to pick up on the field of vibrating information will be able to see the possibilities of the future. As the theory of superposition says, all possibilities exist simultaneously: it is only when we consciously observe that our reality collapses to one possible outcome. It follows that this information is constantly buzzing around the ocean of energy, ready for you to tune into it if you can perceive it. Psychic people are merely those that are highly skilled or naturally open to tuning into this vast field of holographic information.

4. **What is all this talk about auras, chakras, and the like?**

Now, before you decide to throw this book down and say, "I'm not reading any hippie crap," please stay with me. Forget the original frequency you associated with words like *chakra* and *aura* for the time being. You can go back to that belief if you choose to later, but for the sake of the examples in this chapter, please stay with me for now. And trust me: I was definitely one of those people who didn't want to turn into a tree hugger who loved the air and bunnies, wore multicolor woolen jackets, and hummed for world peace. This is why I wrote this book—to show the science behind some real, beneficial ideas that can help all human beings to grow as individuals, as a race, and as a global consciousness.

When we talk about auras and chakras, we are talking about those subtle bodies that are a part of your holosoma or group of bodies. The energetic body is the most immediate body and the closest to our own frequency, which is why there are a lot of people that can see or feel auras. In fact, science has reached a point where it can start to measure or even photograph these kinds of bodies. "As the state of the art of our scientific equipment becomes more sophisticated, we are able to measure finer qualities of the UEF (Universal Energy Field). From these investigations we can surmise that the UEF is probably composed of an energy previously undefined by Western science, or possibly a matter of a finer substance than we generally considered matter to be. If we define matter as condensed energy, the UEF may exist between the presently considered realm of matter and that of energy."[117]

Barbara Ann Brennan, who was a research scientist for NASA and is now a practicing healer, psychotherapist, and scientist, believes "the phenomenon of the aura is clearly both inside and outside linear time and three-dimensional space ... From the holographic framework of reality each piece of the aura not only represents, but also contains, the whole."[118] This evidence gives way to an understanding of why some clairvoyants or psychics can look at your aura and pick up your information. Hovering in your aura is the holographic information of your past, present, and future. Tuning into any one of these vibrating particles can show all past information about you and the superpositioned possibilities of your future. Those that can tune into these frequencies are able to perceive this information.

It's all well and good that I can show you the science of auras, but I bet your experiences will mean more. You may have felt your own or other people's auras before without knowing that you were doing so. Ted Andrews talks about several things you might identify with as ways of experiencing other people's energy fields in his book *How to See and Read the Aura*:

- Have you ever felt when someone is staring at you?
- Have you ever been able to sense how someone is feeling, in spite of how this person was acting?
- Have you ever been able to sense another person's presence before you actually heard or saw this person?
- Have you ever ignored or shoved aside a first impression of someone, only to find that it bears itself out eventually?
- Are some rooms more comfortable and enjoyable to be in than others? Do you notice the difference in one room from the next? Did you ever notice how your brother's/sister's room feels different from yours? How about your parents' or children's?[119]

I am sure you have experienced at least one of the things above. It is pretty fascinating when you start to become aware of these feelings. I have also heard and read that you can actually learn to see people's energy fields, or auras. According to Ted Andrews, "It is believed that you only use about 15 to 20 percent of the cones and rods within your eyes, so it is no wonder that most of us do not detect the subtle light energy of the aura."[120] Ted also explains that you can use "eye charts to exercise the muscles, particularly the iris, you learn to adjust the amount of light you allow through your pupil to the retina. This ability can be developed to the point where you can discern subtle light emanations that you do not normally perceive."[121] Doing the exercises in Ted's book is about increasing the range of frequencies you can see. So like training your muscles in the gym, you can train your eyes to see auras.

Chakras are also part of this energy body. "Like acupoints, chakras are specialized energy centers throughout our bodies where a unique form of subtle environment (life) energy is absorbed and distributed to our cells, organs, and body tissue. However, the flow of this subtle energy (called *prana* by the Hindus) through our chakras is strongly affected by our personality structure and by our emotions, as well as by our state of spiritual development."[122] So essentially chakras are where the ocean of energy connects and flows with our own energy field or energy body. And again, our consciousness, with its thoughts, beliefs, and emotions, can affect this flow:

> It seems anything that causes a blockage or disturbance in the flow of subtle energy through one or more chakras can also lead to the development of illness in the body.
> One of the most important causes of chakra blockage is chronic emotional stress and emotional energy imbalance ... if our thoughts and our emotions become imbalanced because of chronic emotional stress, a constricted flow of life energy through one or more of the seven major chakras can result. This constricted life-energy flow may then produce a weakness or predisposition to illness in different areas of the body.[123]

Caroline Myss, through her research as a teacher and medical intuitive, describes the location of the seven chakras and their emotional attributes:[124]

- Tribe—located at the base of the spine
- Physical Power—located in the genital region
- Self—located in the solar plexus
- Love—located in the heart region
- Will—located in the throat
- Mind—located in the forehead
- Oneness—located at the crown of the head[125]

Myss explains that the main way to keep your energy body flowing and free of disturbances that cause illness is by staying in the present moment. "To promote the free flow of energy through your body, you must focus your attention in the present. Don't carry around thoughts or perceptions that drain your power."[126] I will talk further about health and healing under the subheading Miracle Healings in the Mind Over Matter section of this chapter.

5. Are OBEs and NDEs possible, and if they are, why doesn't everyone experience them?

Scientific research on OBEs and NDEs is done all over the world, and stories and evidence come back in numbers. In *The Holographic Universe,* Michael Talbot writes,

> In the 1960s, Celia Green, the director of the Institute of Psychophysical Research in Oxford, polled 115 students at Southampton University and found that 19 percent admitted to having an OBE. When 380 Oxford students were similarly questioned, 34 percent answered in the affirmative. In a survey of 902 adults Haraldson found that 8 percent had experienced being out of their bodies at least once in their life. And a 1980 survey conducted by Dr. Harvey Irwin at the University of New England in Australia revealed that 20 percent of 177 students had experienced an OBE ... the fact remains: OBEs are far more common than people realize.[127]

Talbot continues, "OBEs have also been documented in the lab. In one experiment, parapsychologist Charles Tart was able to get a skilled OBEer he identifies only as Miss Z to identify correctly a five-digit number written on a piece of paper that could only be reached if she were floating in the out-of-body state."[128]

In the area of NDEs, quite a few reports come back with stunning results. Kenneth Ring, a recognized authority on NDEs, cites the following incident in his book *Life at Death:* "as a female NDEer found herself moving through the tunnel and approaching the realm of light, she saw a friend of hers coming back! As they passed, the friend telepathically communicated to her that he had died, but was being 'sent back.' The woman, too, was eventually 'sent back' and after she recovered she discovered that her friend had suffered a cardiac arrest at approximately the same time of her own experience."[129]

Another example "reports a case in which a woman left her body during surgery, floated into the waiting room, and saw that her daughter was wearing mismatched plaids. As it turned out, the maid had dressed the little girl so hastily she had not noticed the error and was astounded when the mother, who did not physically see the little girl that day, commented on the fact."[130]

So we know they are possible—the evidence is there. You may even have experienced one characteristic of an OBE. Have you ever almost fallen asleep and then suddenly felt yourself dropping as you shuddered awake? Many of us sleep in our emotional bodies, about six inches above our physical bodies. So when something like a sound or a feeling causes a disturbance to our physical bodies, we fall back into them. "Many if not all living human beings have a Second Body. For reasons yet unknown, many if not all human beings temporarily separate from their physical bodies via the Second Body during sleep. This is done without conscious memory, except in rare instances. Far more rare are those instances when separation is obtained with conscious intent."[131]

Michael Talbot expresses his belief that, like those who have OBEs, people who have NDEs are experiencing something real: "People who have NDEs are not suffering from hallucinations of delusional fantasies, *but are actually making visits to an entirely different level of reality*".[132] So using the previous breakdown of our new paradigm of the holosoma, we can deduce that when we are having an OBE or NDE, our consciousness is using our emotional body as its vehicle during sleep or death. It remains in contact with our physical body through our energy body. The energy body is like the glue that holds the two together, and only when a full physical death occurs will this energy body release the emotional body from the physical body.

Dr. Edgar Mitchell confirms the statements made by Talbot and Monroe in his book *Psychic Exploration: A Challenge for Science:* "Man has a non-physical soul of some sort that is capable, under certain conditions, of leaving the physical body. This soul, as manifested in what we call the second body, is the seat of consciousness. While it is like an ordinary physical body in some ways, it is not subject to most of the physical laws of space and time and so is able to travel about at will."[133] So when you experience an OBE, you are actually in the dimension called Locale II or the *paratropospheric* dimension. You can move around by pure thought control, since your consciousness, as in our dimension, affects the reality of that dimension. The main difference is that the OBE dimension is subtler and quicker in responding to your thoughts and emotions.

> Access to holographic reality becomes experientially available when one's consciousness is freed from its dependence on the physical body. So long as one remains tied to the body and its sensory modalities, holographic reality at best can only be an intellectual construct. When one [is freed from the body] one experiences it directly.

> That is why mystics speak about their visions with such certitude and conviction, while those who haven't experienced this realm for themselves are left feeling skeptical or even indifferent.[134]

OBEs and NDEs, like ghosts, are possible. In fact, they are two of the same thing. The only difference is that one is still attached to the physical body, while the other is purely extraphysical.

MIND INFLUENCING MATTER

1. **How can our beliefs change our reality—or even be a matter of life and death?**

At the beginning of the book, I spoke about the aboriginal cultural belief called the pointing of the bones. That is, if a member of the tribe broke the law and the elders decided that it was a bad enough sin, they would punish the person by pointing the bones at him. This would ultimately mean death for the guilty party, and death did indeed occur as a result of this ceremony. The source of this seemingly magical power of the bones has now been uncovered. It isn't that the bones were magical; it was the intentions of the elders as well as the intention of the victim to die that gave this ceremony its power. In chapter 2, I illustrated the double-slit experiment, which was one of the first experiments to confirm the notion that an observer, or consciousness, can affect the result. The famous PEAR studies are prime examples of people affecting the results of experiments purely through their intentions. "On the most profound level, the PEAR studies also suggest that reality is created by each of us *only by our attention.* At the lowest level of mind and matter, each of us creates the world".[135]

So going by this notion, we should be far more observant about what we pay attention to. If you find that you're paying more attention to the negatives in your life, the reality that will subsequently be created will resonate with that concentration of thought. On the other hand, if you focus on expressing gratitude for all the good things that come into your life, your reality will increasingly replicate your state of mind. Wherever you put your attention, your reality will follow.

If you'd like to experience the power of your attention to change your reality, try this. First, you need to be in a room or place you haven't been before or don't know well. The first part of this experiment is to focus really hard as you look around this room, paying *all* your attention to everything in the room that is blue. So take a few minutes to do that before reading on. Once you have done that, keep your eyes on this page. Now I want you to try to list all the brown things in the room. No peeking—keep your eyes on the book. Can you think of anything brown near any of the blue stuff you focused on? Probably not. This is exactly what this experiment illustrates to you: if you keep your attention on all the negative things in your life, it will be almost impossible for you to see the positives even if they are staring you in the face! Your reality will completely shift, and you will resonate with everything negative. The tuning forks of positivity will remain silent and invisible because you aren't resonating at that frequency.

The PEAR studies illustrated a few more profound facts. "Jahn had proved that, at least on the subatomic level, there was such a thing as mind over matter. But he'd demonstrated something even more fundamental about the powerful nature of human intention. The REG data offered a tiny window into the very essence of human creativity—its capacity to create, to organize, even to heal."[136]

The power of mind over matter exists—it has been proven. The story I told you about the guy who froze to death in a truck freezer when it was off demonstrates how someone could create a reality around himself that actually caused him to freeze. What is interesting is that when I told the story to my father, he told me a different story he had heard that was the complete opposite. This story was about a guy who got stuck in a freezer that was actually on but who survived over the weekend by believing that he would walk out of the truck alive. His reality adjusted itself to mirror his belief, as he found anything and everything in that truck that would keep him alive. He moved things back and forth to keep his body heat up and just focused on his very strong belief that he would stay alive.

Your beliefs about this universe are able, at the extremes, to cause life or death. On an everyday scale, they can change your reality to one that works with you or one that works completely against you. It's your choice! You are in charge of your beliefs, so it's up to you to change them if they don't serve you. Limiting beliefs cause a limited reality. Mitchell says in *The Way of the Explorer* that "our beliefs are our map of reality. We do not perceive reality directly, but only the information our senses present to the brain at any given moment, which is then compared with the existing remembered experiences to obtain meaning. Because this map is the only reality we humans know, we often make the mistake of thinking that our map is reality itself, when in fact it is just an incomplete portrait painted from memory."[137]

Throughout our lives we have built up a stack of beliefs about reality. Some serve us—like the notion that running across a busy road is dangerous. Others don't: "I am hopeless, I can't do anything right, I always forget things, I don't deserve to be successful." These things don't help anyone to move forward. As Mitchell says, "beliefs accumulate moment by moment and experience by experience throughout our lifetime."[138] Fortunately, these beliefs can change in an instant if we want them to, or our egos can defend them for fear of being wrong. However, what is worth more—beliefs that help you move easily through the success of your reality, or beliefs that hold on to a false sense of integrity? Your beliefs may not be as life threatening as the one adopted by Paul, but you may as well treat them that way. Because if you are being held back instead of moving forward, you are in a sort of limbo equal to death. So live your life, and find beliefs that serve you.

2. **What can we learn from autism about the possibilities of the mind?**

Previously, I spoke about autism and the amazing ability autistic people have to harness the mind in ways that the average Joe Blow seems unable to. I spoke about how there seems to be some kind of filter that autistic people bypass to get to the information that we restrict ourselves from obtaining. I learned this from my mother, who told me about Dr. Allan Snyder of the Centre for the Mind at the University of Sydney. An article written in *The Economist* on April 16, 2009, titled "Autism and Extraordinary Ability" talks about what he discovered:

> There are examples of people suddenly developing extraordinary skills in painting and music in adult life as a result of brain damage caused by accidents or strokes. That, perhaps, is too high a price to pay. But Allan Snyder of the University of Sydney has been able to induce what looks like a temporary version of this phenomenon using magnetism.
>
> Dr Snyder argues that savant skills are latent in everyone, but that access to them is inhibited in non-savants by other neurological processes. He is able to remove this inhibition using a technique called repetitive transcranial magnetic stimulation. Applying a magnetic field to part of the brain disrupts the electrical activity of the nerve cells for a few seconds. Applying such a field repeatedly can have effects that last for an hour or so.
>
> The technique has been approved for the treatment of depression, and is being tested against several other conditions, including Parkinson's disease and migraines. Dr Snyder, however, has found that stimulating an area called the left anterior temporal lobe improves people's ability to draw things like animals and faces from memory. It helps them, too, with other tasks savants do famously

> well—proofreading, for example, and estimating the number of objects in a large group, such as a pile of match sticks.[139]

Perhaps what Dr. Snyder is doing through magnetism is actually connecting the brain with the vast pool of information available in the ocean of energy, thereby enabling patients to pluck from the ether the information and skills necessary to perform at a savant level.

I find this experiment interesting, yet funnily enough, I think it could be restricting. On the one hand, I can see that it shows people that our savant abilities are ready and waiting for anyone of us to tap into. On the other hand, though, people may view this on a physical level and assume that in order to produce the desired results, you need a magnetic field to disrupt the electrical activity of the nerve cells in the brain.

I want to take this further. If we go back to my theories about energy, we realize that we are all made of energy. Magnetic fields and electric pulses are just another form of energy. We have learned that thought can control energy vibrations, so why not conclude that anyone has the ability to tap into their savant skills—that anyone, if they practice enough, can access this ocean of holographic information? If you believe strongly enough in your own abilities and feel that you have amazing skills, eventually your reality will change the neurological and energetic structure of your brain to help you tap into that.

I came across another interesting concept while reading about Blake Cochran, an autistic boy who ended up becoming a normal boy. As holistic medical practitioner Dr. Karen Pryor treated Blake throughout his journey, she realized that his autism and introverted behavior were the result of his ability to see and hear more dimensions than most people. "Karen and Jett agreed that Blake might be one individual who can see and hear in … a wider frequency spectrum than most. They explain why people continually asked Blake, 'What are you doing?' or, 'Why are you talking to space?' Karen continues, 'In reality, we are the limited beings. Blake is able to receive the information from a computer screen or a television.'"[140] They ended up being able to awaken him out of this state by addressing their teaching methods across this spectrum all at once. To any one of us, learning this way would be overwhelming, but for Blake it clicked. He could pay attention, and slowly he came out of his shell. This is just another illustration of how autism gives a sneak peak into the possibilities of our currently limited minds.

In another area we talked about, past life recall, Amit Goswami talks about an autistic girl who was successfully brought out of her withdrawn state by allowing her to address her past life:

> An autistic five-year-old was brought to Helen Wambach, a clinical psychologist. This child, Linda, was severely withdrawn and refused all contact with the therapist until through role-playing she was allowed to repeatedly force-feed her therapist from a baby bottle. Now Linda was able to reveal how much she hated the helplessness of infancy. Contact was now established and rapid progress followed, and soon Linda was like any other five-year-old.
> Now what is also interesting in this case is that, as an autistic child, Linda possessed high math and reading skills, skills she lost when she became normal. Wambach says that Linda's autistic behavior was due to the child's holding on to the adult identity of a previous life. When she came to accept her new condition of a child with the help of her therapist, she gave up her adult identity and lost her adult skills (Wambach 1978).[141]

Autism is unique. In every case, autistic people are different from the norm to varying degrees. Autism is a very unfortunate condition and extremely tough on parents. However, the behaviors of children with the condition are furnishing us with interesting strands of information, and perhaps they can provide evidential support for some of the theories I have talked about in this book.

3. How do miraculous healings work, and why don't they work for everyone?

Carolyn Myss, Dr. Bruce Lipton, Barbara Ann Brennan, Dr. Joe Dispenza, and Inna Segal are scientists or practitioners that have researched the connection between energy and our physical health. They have all come to the conclusion that our beliefs affect our energy, which in turn has an effect on our physical bodies. Think of Dr. Masaru Emoto and his experiments with water crystals—how the energy of thought affected the water. We are 70 percent water and 99 percent energy particles and only 1 percent matter. The method applied by Dr. Emoto will absolutely have the same results on our bodies.

Biologist Bruce Lipton has discovered that our cells respond to energetic perceptions rather than, say, an electrical response.[142] Our cells have millions of jobs to do. What tells them to do their jobs is a signal. If our cells need to increase our body temperature because we are in a cold environment, then our senses that feel the cold send them a signal, and they create heat. Okay, now think back to Paul in the freezer. His thoughts actually overrode his sense for cold and told his cells to start feeling the effects of hypothermia. So it is actually our thoughts and beliefs that control our cells. We can use this for positive change. We can control our health on a cellular level.

I am a person who never gets sick. My mum said that to me once, and I decided from a young age to live by that. I don't know why, but it seemed like a pretty positive thing I could be proud of. And to this day, I truly don't get sick. There have been a few occasions—say, twice I can think of in the last five years—when my mind has been so stressed that eventually the power of the thought "I never get sick" was weakened and I did get sick. If anything, it was my body saying, "If you don't stop and have a break, I will." But I suspect if you examined someone who got sick all the time along with someone who never did, that one immune system would be stronger than the other. This might be the point where you say that they were probably born with that immune system—it's genetics. Well, what if I told you that genetics no longer have anything to do with what we can and can't do?

Dr. Bruce Lipton talks about how scientists have discovered that our DNA is actually only a blueprint for who we are. It is the power of our minds—our beliefs and our subconscious—that decides which parts of the DNA to activate at any time. Now doesn't that put a whole new perspective on it?[143]

So if Dr. Lipton has helped us to see that our beliefs can change our biology, how does it work? How do our thoughts truly affect our physical bodies?

Well, it has to do with the energy centers that I talked about previously—our chakras, which indirectly affect our physical bodies. Carolyn Myss explains that "When the free flow of the life force is impeded within the body's energetic system, illness can develop. Imagine your body as an intricate database. Every thought, every feeling, every memory you ever had has been encoded and converted into matter—a form of cellular memory."[144] Depending on what type of negative thoughts you have, they affect different chakras, which in turn affect specific areas of the body. For example, Carolyn Myss says that "experiences concerning your self-esteem are logged in the third chakra [the solar plexus; associated with the self or ego]."[145]

The following personal example of how an issue can affect you on a physical level did not become obvious until this year. It took me ten years to resolve my negative emotional attachment to my dad having an affair. I thought I had forgiven him, but subconsciously, and more importantly, energetically, I was still holding onto some limiting thoughts. I had a lot of issues with guys. I couldn't truly trust them, because I believed that even the nice ones might cheat on me. Emotionally, I was feeding the belief that if my dad was a nice guy, how would I ever find a nice guy that wouldn't cheat on me? This was a limiting belief that I put on myself, and it actually had nothing to do with what happened.

Eventually, I went to two separate energy healers—one who worked with *prana* (the Hindu idea of the life force inherent in all things) and another who could see auras. Both of them could sense that my chakra in my base and genital regions had been affected, that energy was being drained from that area, and that my chakras were depleted. The gentleman who could see my auras sensed that something had happened to me at age twenty-three. When I worked out where I was at twenty-three, I realized that it was the year I found out my dad had had an affair. The healer had obviously been able to access the holographic information in the field around my body.

In the meantime, I also visited a doctor, and it was around that time that I found out that I had CIN 3 cells in my cervix, which is where my chakra was showing negative energy. I found this coincidence quite amazing and highly significant. I decided to rid myself of my negative beliefs and work through a couple of different activities that both healers suggested as an aid in assisting me to let go of the limiting thoughts that were holding onto the negative past. Once this changed, I could feel that my energy had changed. I had the CIN 3 cells cut out and have since then been given the all clear! I believe that they won't come back again, because of the change in my energy. I feel that this may be where a lot of cancers come from—subconscious, limiting beliefs or negativity held onto from past events.

I'm not surprised that physical medical science finds it difficult to find a cure for cancer, because I believe it stems from thoughts that affect our energy and, when not treated, affect us physically. The medical system today thinks "only about those matters they can touch or taste or feel with their physical five senses. And yet it is this realm of spirit that is the hidden reality behind the true nature of health and illness when viewed from the perspective of vibrational medicine and its understanding of the multidimensional human being."[146] I believe that we need to cure cancer from the level of consciousness—the level that deals with our thoughts, beliefs, and emotions—in conjunction with energy healing and healthy eating and living. Author and publisher Louise Hay famously did this and writes about it in her book *You Can Heal Yourself.*

Neuroscientist and chiropractor Joe Dispenza talks about an example of this on his audio CD *Evolve Your Brain: The Science of Changing Your Mind.*[147] An experiment was done on people with diabetes. There were two groups of people with type 2 medically dependent diabetes. They had their blood sugar measured and then watched either a humorous lecture or a one-hour comedy show. They were then fed a meal, after which they had their blood sugar measured. Those that watched the humorous lecture had an average blood-sugar level of 123 mg/deciliter (70–120 is normal). Those that watched the comedy achieved 77 mg/deciliter. No insulin was needed—it was pure joy that healed them. After this study, the scientists discovered that they had turned on twenty-four genes just by being happy. This positive energy was all that helped them heal.

Another absolutely mind-blowing video I saw included some footage of a lady's bladder cancer being healed right in front of my eyes on an ultrasound machine. In Gregg Braden's interview in *The Science of Miracles,*[148] he talks about the power of intention—of surrounding yourself with positive vibrations and, most importantly, of putting yourself into the frequency of being healthy. The video shows a lady in China who goes to an energy healing center to cure her bladder cancer. She and three practitioners chant a word that to them represents that she is completely healthy. The chanted word sends a strong positive frequency to the cancer, which is completely depleted within three minutes right before our very eyes. It is absolutely paradigm-shifting. You can see it to believe it!

The placebo effect is another interesting example of the power of mind over matter when it comes to healing. Talbot recounts an interesting story about this from California:

> Dr. David Sobel, a placebo specialist at Kaiser Hospital, California, relates the story of a doctor treating an asthma patient who was having an unusually difficult time keeping his bronchial tubes open. The doctor ordered a sample of a potent new medicine from a pharmaceutical company and gave it to the man. Within minutes the man showed spectacular improvement and breathed more

> easily. However, the next time he had an attack, the doctor decided to see what would happen if he gave the man a placebo. This time the man complained that there must be something wrong with the prescription because it didn't completely eliminate his breathing difficulty. This convinced the doctor that the sample drug was indeed a potent new asthma medication—until he received a letter from the pharmaceutical company informing him that instead of the new drug, they had accidentally sent him a placebo![149]

Talbot continues, "Understanding the role such factors play in a placebo's effectiveness is important, for it shows how our ability to control the body holographic is molded by our beliefs. Our minds have the power to get rid of warts, to clear our bronchial tubes, and to mimic the painkilling ability of morphine, but because we are unaware that we possess the power, we must be fooled into using it."[150] Which brings me to the point of why it doesn't work for everyone: it is simply because of their beliefs. If they don't believe it, they won't see the effect. If they believe they will only get better with medication, that is exactly what will happen. It is purely belief that does or doesn't allow for miraculous healings. Beliefs are powerful things!

4. **Is the law of attraction real, and if so, how does it work?**

The law of attraction is real in a sense. I don't like to call it the law of attraction, simply because I don't see that it is an attraction per se. It's not like when a magnet sticks to a fridge. It's not that kind of attraction. "Attraction is an idea that comes from the linear mind-set of separation: what you want is 'over there,' and you must make it come to you with will and cleverness. In a unified world, instead of attracting something, you notice it *already existing in your reality,* and you keep paying attention to it. No effort required".[151] It's more about tuning into the reality that you want. It has more to do with resonance. So when you're vibrating at a frequency of happiness, your reality will tune to that happiness. Your reality will only resonate with things that are of the happy frequency. Think of the tuning fork example. There are a hundred tuning forks with different frequencies, and as soon as you hit one frequency, all the other tuning forks at the same frequency will start resonating with it. It is exactly the same in your life. You will only see the things that you resonate with. So if you're tuned to happiness, you will only tune into things that you are happy with. As

Penney Peirce says, "No willpower is needed to 'get' what you want because the next just-right thing is always showing up."[152]

It's not like you try to attract a BMW, and then all of a sudden a BMW comes dragging itself out of a car yard and flies and skids over to you because you're just so magnetic. It will begin to resonate in your reality because you have resonated to the frequency of that BMW. All of a sudden, things will happen. Synchronicities happen because the resonance is there. The BMW resonates with your bank account, and your bank account resonates with a car salesman. It sounds funny, but that's exactly what's happening. So it's not so much a law of attraction—it's more a case of tuning into the resonance of your reality. Gregg Braden puts it brilliantly: he says that our lives are a mirror. However you are, your reality will mirror it. If you're depressed, your reality will mirror that and show you a depressed reality. It all has to do with the frequency you're on and what else in your environment is on that frequency: "The man or woman who is filled with Love sees Love on all sides and attracts the Love of others. The man with Hate in his heart gets all the Hate he can stand. The man who thinks Fight generally runs up against all the Fight he wants before he gets through. And so it goes, each gets what he calls for over the wireless telegraphy of the Mind."[153]

It is also very important to understand the real intention behind what it is you are trying to resonate with. If you are trying to resonate with wealth because you are afraid of being poor, then it is actually that emotional tie with fear that will resonate more than the thought of wealth. To tune into the *real* purpose, the true positive intention, look closely at why you wish to be wealthy. Ultimately, it usually has to do with being happy. If that is your true intention, resonate with that, and the wealth will come. The ultimate frequency of anyone's reality should be happiness. The more you tune into this frequency, the more it will resonate with your reality.

For me, it isn't necessarily the law of attraction. It is more reality through resonance.

RELIGION AND SCIENCE

> *Science without religion is lame, and religion without science is blind.*[154]
> —Albert Einstein

1. **What about the dichotomy between science and religion?**

To be perfectly honest, I tend to agree with Einstein above. Both science and religion are important, and after all the research I have done, I feel that science proves why religion works so well. At the beginning of the book, I talked about the majority of religions having four main characteristics: a god, beliefs, faith, and prayer. It was through studying these components that I discovered that science and religion were much the same. Dr. Edgar Mitchell feels the same way. In *The Way of the Explorer,* he says that "parallels emerge from the data, similar patterns of structure between religions and this new brand of physics."[155]

God Most religions have some sort of omnipresent being that has control over everything. Whether the god is within or an external being, it has the ability to be everywhere at once, to be nowhere, to listen to you, and to answer your prayers. He/She contains all the answers of the universe. This God created the world and can destroy it. Well, science has found God right under our noses: the ocean of entangled holographic energy that makes up our own selves and everything else and that completely manipulates itself according to our "prayers." Whether you want to envisage God as a being or person of some sort, the facts remain the same. It listens to you when you're in need, it is ever-present, and it will give you what you want. If you are in a positive frequency, it will give you this reality. If you are negative or bad, it will give you the equivalent. It will make your life heaven or hell according to how you behave, what you believe, and how you think. God is our forever-knowing ocean of energy.

Beliefs I have already discussed beliefs in some depth. Whether your beliefs have to do with science or religion, they create the structure and limitations of your reality. Beliefs are the foundation of a religion, just as they are the foundation of the reality you've created according to your beliefs about science. When you consciously accept and follow the beliefs of your religion, the wave of possibilities will collapse accordingly to support your beliefs. It is up to you to find the belief structure that resonates with you and serves you according to your Ultimate Frequency of Happiness.

Faith Faith means having absolutely no doubt that your beliefs are how the universe is. It means hoping and trusting in your religion and your reality. This kind of faith is on an energetic level, and it connects your frequency with your thoughts. If you have faith that your beliefs are true for you, then this feeling will ultimately shift your frequency, which will then resonate in your reality. If you have lost hope, then the effect of your thoughts and beliefs will get lost in the translation, and your reality will reflect this loss. You will start to see a reality that doesn't conform to your beliefs, thereby making you lose more faith. It is a lose-lose situation. Once you realize the important role of faith in the equation, whatever you believe will be your reality. Belief without faith will create a reality that doesn't quite reflect your beliefs. However, having unequivocal faith in your belief confirms your reality and therefore your beliefs.

Prayer Prayer is the foundation for miracles, both religiously and scientifically. Prayer in the religious sense is about direct communication with your maker. It is the direct line that allows you to put in your order for the wants and wishes in your life. Religions recognize the power of prayer, and science is only now realizing why. But the key to prayer is not asking or wishing for what you want. It is being in the frequency of already having what you want. It is putting your attention onto this frequency to allow the ocean of entangled holographic energy to resonate and respond.

Gregg Braden talks about an experience that illustrates this in his documentary *The Science of Miracles*. He tells us of a Native American man who wanted to pray for rain because there had been a drought in Indiana for some time. This man took Gregg with him, and they trekked to an ancient ruin. When they got there, the Native American began a small ritual by taking off his shoes and stepping into the circle. He closed his eyes, put his hands together, and then prayed. He only did this for a few moments and then turned around, ready to go. Gregg, surprised that it wasn't a much longer proceeding, asked him what he had prayed for. The Native American man replied that it wasn't about what to pray for, because to do that would be to inadvertently affirm the absence of rain (you would be sending that frequency out). Instead, he had visualized what it was like in his village with the rain coming down—what it felt like with the water and mud in his toes, what the fresh rain smelled like, what sounds surrounded him, what it looked like. He was feeling the frequency of the rain. Braden said that they watched the weather on TV that night and that clouds literally seemed to have appeared from nowhere. It rained later that evening and for the next few days.

Whether you are religious or not, prayer is a significant aid to creating the reality you want.

Once you realize that the core components of religion and science are very similar, then you might almost think that the context of each religion doesn't matter as long as you follow those four points. Well, I thought that at first. I came to realize, however, that a huge part of religion is about individuals. Each of us is a highly unique consciousness vibrating at an infinite range of frequencies. Just as each one of us is unique, religions, too, have their unique qualities. One religion may not resonate with you while another complements you to the core. Although the context is not the most important part, it is important as a way of providing variety for the massive variety of individuals that populate our planet. What is most important, though, is to have respect for your God, to have a positive and supportive belief structure, to be completely and utterly faithful to your beliefs, and to pray constantly to keep on track with your reality.

It is a bit of a catch-22, this idea, because if you get distracted by the notion that your belief isn't valid because someone else's belief contradicts it, then you will lose faith. This is how fights, conflicts, and wars begin—the ego's battle to believe in the "right" religion. When you come from a space where you are responsible for your own reality, then you realize that you have to believe it to see it—you realize that for your religion to be the right religion for you, you have to have complete faith in it. Your reality will reflect this. Everyone can be right. Essentially, if you mind your own reality, everyone else's reality won't affect yours.

Think about it this way: if you focus on your reality, on remaining in a positive frequency of happiness, only things that resonate on that level will be a part of your reality. If we all did this and remained at that happiness frequency, those that achieved it would see peace in the world and nothing else. Those that didn't would only see the conflict in the world. If you want to see peace in the world, you have to first create it in your reality. There is a saying attributed to Mahatma Gandhi that expresses it best: "Be the change you want to see in the world." With the background of this scientific research, this quote suddenly has more significance. If you want to resonate only with the tuning forks that resonate peace, you first have to resonate at that frequency yourself to become a part of it.

	The Summary of Religion & Science as One
Religion	*Science*
God	The ocean of entangled holographic energy, the fabric of this universe
Beliefs	Conscious influence over reality
Faith	The emotion and intention behind the thought waves (being open/expecting to receive)
Prayer	Connecting with or tuning into the frequency of the ocean of energy

2. **Why is religion so powerful and important in people's lives, and is there a right or wrong one?**

Did you ever play telephone as a kid? (Stay with me—there is a method to my madness!) It's that game where someone starts with a sentence. It's whispered into one person's ear, and then another, until it goes around in a circle and comes to the last person. The last person says out loud what they believe they heard, and then the first person usually reveals something totally different! Well, wait for it—here's the segue! What if the original sentence were a story about how everything worked in the universe, how to live life fully, and how to create the reality of your dreams? Then this story was told or whispered into different people's ears through different languages, generations, and so on. This is how I believe we got a complete array of religions, all with similar characteristics: a God, beliefs, faith, and prayer.

Religion is very important, since it gives guidance to those who need some guidance with their beliefs. It gives them a foundation from which to create the reality they want. Without a connection to the ocean of entangled holographic energy (God) and without a way of creating our reality (beliefs, faith, and prayer), we would be lost in the rips, swirls, and tides of everyone else's reality. We need religion or science to provide that structure so that we can be responsible for our own reality and connect with our own frequency of happiness.

The only right religion is the one that resonates with you and helps you to achieve the ultimate reality for you. No religion is better than the others; you just need to find what helps you to be the best you can be and treats every unique individual's reality with equal respect.

3. **Are superstitions real?**

Superstitions are beliefs that have been passed down through the generations and that aren't necessarily connected with any religion or culture—like the number thirteen being bad luck for some and good luck for others. What I think I have shown so far throughout this chapter is that the actual substance of a belief does not necessarily have to be factual. It's the energy behind the belief that gives it power and can bring it to life. Say, for example, that you utterly believe that if a black cat walks across your path, you will have bad luck. You collapse any other possibilities in the field of energy from your thought frequencies, and that creates this reality. So later you may trip or lose some money, and you will then associate that with this superstition. It's not just that you will relate it back to this belief: the frequency that you sent out made other bad luck tuning forks resonate in your reality, so that is what you notice. Superstitions are only real if you put the power of belief behind them. Take this away, and they are nothing but a bunch of words in a sentence.

4. **How does Murphy's Law work?**

Murphy's Law is an amusing one, because it shows how everything actually works, but in reverse. What's funny is that if people just believed in the opposite of Murphy's Law, their realities would reflect exactly what they wanted—not what Murphy wanted. Jones writes that "Some parapsychologists even believe that Murphy's Law—everything that can go wrong will go wrong—may also have scientific basis, suggesting that energy attracts like energy and that the human mind can indeed interact with inanimate objects and cause them to fail, usually at the most inopportune of times."[156]

Murphy's Law is just like that example from the William Braud experiment with the REG/RNG machines, in which he illustrated that if people didn't believe in the experiment, their lack of faith would affect the outcome. They wouldn't be able to affect the results in a positive way.

However, their attitudes actually caused results in the opposite direction, illustrating that the experiment was, in fact, working.

Murphy's Law is the same thing. If people believe 100 percent that Murphy's Law is going to apply to them that day, it will, and I myself have experienced this. If I am in the Murphy's Law mode, I know that if I want it to stay sunny, I will forget my sunglasses on purpose. Inevitably, it will stay sunny. This is because I have actually connected to the frequency of squinting in the sun and being uncomfortable. I mean, this doesn't happen all the time, because you don't always consciously think about Murphy's Law, but once you start thinking about it, that becomes your reality. Thoughts affect the feelings that affect the energy particles that affect your reality. So Murphy's Law is a great illustration. If you've ever experienced Murphy's Law, then that is a pure example of your observations affecting reality. Whatever you didn't want to happen happened because all your energy, all your emotion and feeling, were connected to what you didn't want to happen, so that's what collapsed the possibilities into reality.

Whatever you believe will become your reality—whether it's Murphy's Law, superstition, or religion. It all has to do with your beliefs affecting your reality. So now that you have read through the answers and understood the theory, you might be wondering: How do I actually implement this paradigm in my current world? How do I create my reality? The next chapter is dedicated to helping you make the most of this paradigm.

Chapter 4: The Search for Resonance

Well, for those of you who have been wondering, here's how to implement this new paradigm. You've asked yourself the questions or had me ask them for you. You've had the science explained and your questions answered. A whole new paradigm is present in your reality—but how do you apply this new knowledge to your life?

This is what this chapter is about. First, I will help you work out where you are—that is, work out if your fork is in tune. Second, we will work together to find your ultimate frequency and where you want to go in life. Third, we will pull out all the barriers and excuses that normally get in the way when you try to move toward your goals, and I'll provide you with a full tool kit of resources to help you on your journey. Finally, we'll put a strategy in place for you, and you will commit to your new frequency.

A few summarized guidelines at the end of this chapter should help you stay on track. If you are ready to take on the challenge, read on! If your life is perfect and everything is already there or you're perhaps not ready to move toward your new frequency, then skip this chapter. It will only help you if you're ready to take responsibility for your reality and to experiment with this new way of thinking.

IS YOUR FORK IN TUNE?

How are you feeling right now? I suspect you are most likely feeling highly inspired and excited to tell the world about the best book you have ever read! Okay, I may be playing with you a little. An author has to have some fun!

But how are you feeling about life right now overall?

I want you to imagine that you are a tuning fork. As you know, when you vibrate at a particular frequency, other things around you at the same frequency will resonate with you. Resonance is the key indicator in your life's journey. Always be aware of what is resonating with you. For example, let's take today as a snippet of your life—what resonated with you in your life? Was it red traffic lights, stop signs, angry drivers, more work at your job, a bad meeting, people getting annoyed with you? Or was it a straight run to work with green traffic lights, a pay rise, the best lunch ever served, finding five dollars on the ground, meeting some really fun people, or just a relaxed, peaceful day? What in your reality is resonating with you?

Reality

Is your fork in tune with the reality you want, or is it out of sync? If your reality is mirroring everything you've ever wanted, I would say that you're resonating to your ultimate frequency. If it isn't, we need to tune your fork.

Don't forget to break your reality down. You may actually have most of it in tune, yet one area of your life may not be resonating with the best results. Could the issue be in any of these areas: finances, relationships, health, spirituality, career? Perhaps it's something completely different.

First, you need to do a reality check. If something keeps showing up in your reality that you really don't want there, that is the area you need to work on. So what's the next step? Once you have assessed your reality, we need to work backward toward *you*. If you remember chapter 2, the *path of affect* goes like this:

Consciousness → Thoughts and Beliefs → Emotions → Energy → Reality

If something is showing up in your reality, a lot of effort has been put into it for it to actually appear. It has to have been backed by a continuous stream of thoughts and beliefs that affected your emotions enough to have an effect on your energy, which in turn eventually turned the idea into reality. If the reality is a pain in your body, then you need to work backward to undo what has been done. Rather than just popping a Band-Aid on your reality, you need to work at breaking down the pathway of creation to the seed, or cause—your thoughts and beliefs. The reason you can actually break this down is that *you* are the person who put the seed in the ground to begin with. You can take it out as well—you can stop it from affecting your emotions, energy, and reality.

Energy

What is really awesome is that you can even stop bad things from happening in your reality by catching them before they come to fruition by assessing your energy (aura or chakras), becoming aware of and assessing your emotions (including the suppressed ones), and changing your thoughts and beliefs. If you want to get an assessment of your energy, you can either do it yourself or visit someone who has access to a wider range of frequencies and can see or feel your aura or chakra.

If you decide to do it yourself, sit on your own in a quiet room and really let yourself become present with and around your body. Ask yourself at first, How energetic do I feel? Am I bursting with energy, or am I lethargic? This will be your first indicator. From there, go through and analyze your body. Notice any dark spots you sense, any tweaks of pain. These areas may be trying to tell you something. Grab one of the books I mention in the resources section on analyzing your body to help you figure out which thoughts or emotions the pain might be coming from.

If you decide to see an experienced practitioner, you have a choice between pranic healing, energy healing, kinesiology, aura reading, and so on. It is always a good idea to go to someone who has been recommended by someone you trust so you feel confident that you are going to someone who has a genuine ability. Sometimes it's also good to get a couple of opinions from different experts. It's always interesting to see what they pick up, because they will sometimes pick up on something that isn't obvious to you—so always be open to the information. You can take it or leave it once you have made your own assessment.

Once you've received an energy reading, it might back up what is going on in your reality or even point you toward another, more pressing area that you need to work on. When I went to the guy who could read auras and then to the lady who was a pranic healer, it really helped me to concretize what I had suspected from my own observations about my reality. They were even able to provide me with practical things I could do to heal the energy. What is also interesting is that they both worked on assisting me to eliminate some limiting beliefs and thoughts.

Once you've analyzed your energy and pinpointed areas of concern, it's a good idea to check your emotions.

Emotions

> *All physical beings have communication from their inner being in the form of emotion, and so, whenever your emotion is positive, you can know that you are in harmony with your inner intention.* [157]
>
> —Esther Hicks, speaking for Abraham in *The Law of Attraction*

Abraham also talks about using emotions as your thermometer. If you are feeling happy, for example, then you are on the right path. Similarly, anytime anything—especially your thoughts—makes you feel any sort of negativity, even if it's just doubt or a slightly off feeling, be aware that this indicates that you are venturing off your path. You are tuning into a negative frequency.

Emotions are a very important gauge in assisting you to find the ultimate frequency for yourself. If you find that there are consecutive days when you have boundless energy, when you're happy and thrilled with life, then take note of your thoughts, of what is going on—you are most likely resonating at a positive frequency, and this is reflecting itself throughout your reality, energy, emotions, and thoughts. If the opposite is happening, then this is an indicator that something is wrong. When something is wrong with your emotions, the next step is to look at the thoughts that are making you feel this dissonance.

Thoughts and Beliefs

This is where everything begins in your reality, both good and bad. "It's up to you how you want to feel and who you want to be. No one else can create the conditions for you to feel good if you haven't decided to feel good."[158] Too many people rely on everyone else to make them happy, whether it's their partners, families, or friends. The only person who can make you happy, the only person who can make your energy and reality vibrate happiness, is you. You are the center of your own reality, and it's up to you to be responsible for the thoughts and beliefs that you have. "Your personal vibration is generated from within you by your own choices."[159]

When it comes to your thoughts, it's a good idea to write them down in order to help you become aware of what you're thinking. Listen to how you treat yourself. If you make a mistake, do you say, "Oops, glad I did that; now I know what not to do," or do you say, "Stupid idiot, why did you do that?" or "I'm hopeless"? These thoughts will slowly create ingrained, limiting beliefs. They will cause your tuning fork to resonate with similar thoughts and energies. So all you'll resonate with are the things that show you that you're hopeless. What a crappy reality that is!

By now, you should realize that you are the person that has the *choice* to focus on the good or the bad. Yes, you may have created habits, but habits can be broken in an instant if you want to do so. Is your fork in tune? Look around you—what reality do you resonate with? How are your energy levels? Are you boundless or confined? What emotions are you feeling, and most importantly, what thoughts do you think? Write down the areas of your reality that you wish to work on.

Now that you have a full breakdown of where your tuning fork frequency isn't quite right, it's time to re-create it to vibrate at your ultimate frequency. Create the reality you want by redesigning your tuning fork—from thoughts through to reality!

WHAT IS YOUR ULTIMATE FREQUENCY?

You now know where you currently are in your reality. We need to know where you want to go. What is your ultimate frequency? Not everyone knows what this is yet, while some people have known it since the day they were born. It's that feeling of absolute elation, excitement, and contentment. When you are at this frequency, time seems to fly and you can become totally oblivious to what is going on around you. You are so in tune and resonating with what you are doing that you are on a high. The ultimate frequency in life is happiness, and you should be able to get into this frequency regardless of what is going on around you. This is probably one of the biggest challenges in life; however, I believe it is the ultimate meaning to life—to be in the frequency of happiness regardless of your external reality.

As you pursue this goal of attaining your UFH (Ultimate Frequency of Happiness), you will get to a point where nothing ever bothers you. You will resonate so much with your UFH that anything that used to irritate you will be on such a different frequency that you don't even pick it up. It will no longer resonate at your frequency.

If you apply the paradigm we have been talking about throughout the book to the meaning of life, you will notice that your reality always reflects your consciousness. So if you are vibrating at the frequency of happiness, your surroundings will eventually reflect that. If you stay that way for an extended period of time, your life will mirror your state of mind more and more. The more your reality reflects your happiness, the more your happiness will increase. In theory, it has the potential to be a continuous spiral toward an unbelievable level of happiness that can only be understood by those who reach it. It is this unlimited happiness we should aim for rather than material things. Material things provide only limitations on what could be. Penney Pierce puts it well: "if I tried to make my future with my limited mind, it would be a mediocre thing. I knew my life would arise out of the velvety blackness without limitation if I'd just stay in the pleasure, with its positive expectation and trust in goodness."[160]

The ultimate frequency is happiness, and the purpose of life is to attain this vibration regardless of the external environment. Material goals only limit your happiness. "If I have $100,000 in the bank, I will be happy" is a common refrain. The catch is that according to this paradigm, to have

$100,000 in the bank, you have to be happy first; you have to feel the frequency of already having it. It isn't the other way around. Most people eventually find that even though they achieve their ultimate material goals, happiness remains elusive. You have to look within to work out your UFH. Some people travel to find their happiness when it's with them all along. George Moore says, *"A man travels the world over in search of what he needs and returns home to find it."*[161] Why not get a head start on it all and become happy first by resonating with your UFH—then watch the reality reflect your inner vibration?

I am not expecting you to suddenly achieve the UFH straight away; in fact, you may only reach it sporadically at first. It takes baby steps to get there. You have to crawl before you can walk, and you have to walk before you can run. It is the same here. Penney Peirce suggests that "you can improve your overall personal vibration by improving one part of your makeup—just move up the scale to a quality that feels better. For example, if you're exhausted, move into comfort and rest. If you're frustrated, try shifting to contentment. If you're obsessed, try looking at your underlying beliefs."[162]

If you are a person that mostly resonates in a low, depressed frequency—if sad music resonates with you most—then I don't expect you to suddenly switch, to leap to the ultimate frequency and be excessively happy. It takes a gradual increase in your frequency. Don't just jump from being a bass to a soprano: you need to move gradually up the scale. Scuba divers, when they have been deep underwater, have to stop every few minutes to allow their blood to depressurize. It is a similar process here. You need to take it one level at a time and congratulate yourself each time you increase your frequency.

> *The meaning of life is to evolve so that no matter what is going on around you on a multidimensional level, you will be able to attain your Ultimate Frequency of Happiness. The purpose of your life is to find what resonates with you at your Ultimate Frequency of Happiness and to use that to help others reach their highly evolved states of the Ultimate Frequency of Happiness.*

To be able to evolve to this level, you will need to face some hidden negatives deep within your subconscious as well as perhaps some past life memories that contain all the limiting beliefs, thoughts, and emotions that hold you back from maintaining this frequency. I will provide you with tools to help get through those.

Now let's talk about what resonates with you—let's talk about how to find your Ultimate Frequency of Happiness.

Different activities resonate with different people to make them happy. Some people love to play sports; others prefer more creative activities, such as music, drama, or art. Still others prefer to get their hands dirty working with machines, outdoors (as in gardening or farming), or fixing things—human or otherwise. Each of us is unique, and each of us has something that truly resonates with us at the level of that Ultimate Frequency of Happiness. When you're involved in this activity, you lose yourself in whatever you are doing, and time flies because you have such a passion for it. Psychologist Martin Seligman[163] calls it being in a state of flow. He believes that experiencing flow is an indication of authentic happiness, or what I am calling your Ultimate Frequency of Happiness. For Seligman, being in a state of flow requires the following:[164]

- the task is challenging and requires skill
- we concentrate
- there are clear goals
- we get immediate feedback
- we have deep, effortless involvement
- there is a sense of control
- our sense of self vanishes
- time stops

Penney Peirce calls it your *home frequency*: "You have an amazing vibration like this inside you, too—a resonance that conveys your soul's love, truth, abundance, and joy. It bubbles up from your tiny 'quantum entities,' waving out through your cells and tissues to fill the space around you. It's always there, and it remains dependably consistent. This is your *home frequency*, the vibration of your soul as it expresses through your body."[165]

If you want to know what it's like to live a life that resonates with your UFH, let me give you a small comparison. I want you to think about Monday morning. The alarm goes off, and you open one eye and realize it's 6:00 a.m. You have had a good eight hours' sleep, but you have to go to work and feel like you have had no rest. You have no major inspiration. Getting out of bed takes convincing, and only the threat of not being paid eventually pulls you out of bed. This feeling, this constant battle of convincing yourself to do something, is evidence that you are far from your tuning fork frequency! Okay, let's think of another scenario. It's Monday morning, but it's the first day of your vacation. The alarm goes off at 5:00 a.m., and you've had less sleep than normal, but you bound out of bed excited and happy. You are flying overseas on vacation today! You're singing, and you're way too happy at that time of the morning—you can't help but buzz around, as you have boundless energy coming from nowhere.

So how does that happen? Physically, you should be more tired and therefore grumpier. Instead, you have more energy. It's the buzz, the energy vibration that happens when you do something that resonates with you. Imagine if you could have endless energy like this *and* feel that way about your job!

When in your life have you had moments like this? What makes you get out of bed feeling excited in the morning, regardless of how much sleep you have had? Some of you may already know what helps you to be at your UFH. Others may be struggling. It's important to take note of this to help you find what you are meant to be doing in life. It's this being in flow, being at your authentic happiness frequency, that will change your life for the better. When you are in that frequency, your reality will reciprocate. Did you feel it once doing something when you were on vacation? Did you experience it as a child but haven't been back to it since because of a bad experience afterward? It's time to start searching for it and analyzing what it might be. If you still have no idea, then it's time to start trying new things.

Another way of helping you to pinpoint what raises you to your ultimate frequency is a technique that I saw on Oprah. She was interviewing Marcus Cunningham, the author of *Go Put Your Strengths to Work*. The idea was to have a notepad with one side headed "Things I am good at or love doing" and the other "Things I am not good at or hate doing." Take this notebook around with you for a week or so, and when you are doing something you love or hate or when you think of anything significant, just jot it down in those sections. When you are jotting it down, take note of what type of energy you are feeling—whether it's frustration, excitement, satisfaction, or something else.

After a week or so, take a look at everything you have written down. This should start to give you an idea as to which things help give you energy and which things drain you. Once you have a list of things that make you happy—I like to call them energy-building activities—you have a starting point. These are the things that leave you feeling energized after you finish! So put your energy into them, and your mind will send out conscious thoughts to start resonating with more things that bring you that same energy. You will start to notice things coming into your reality that connect you with that frequency.

This journey may take you a month or a couple of years, but as long as you are on the path to finding your ultimate frequency, you have a true purpose and will ultimately stay in a higher frequency. Once you find the activity in life that puts you in your UFH, the next thing is to find out how you can make that your career. The best way is to ask yourself, How can my unique UFH help others? What problems are there in the world that I can solve with my unique frequency? When you answer that, you will have yourself an amazing purpose in life doing something you love.

REALITY CHECK—WHAT CAUSES DISSONANCE?

A lot of you may find that the idea of being in your UFH is so foreign that it seems alien. You may be experiencing the complete opposite, the feeling of dissonance—that your work makes you feel sick inside, that you hate your job but have to do it for the money. This dissonance is a true indication that you are far from your UFH, that you are totally off track from your purpose in life. People use money as an excuse, or they might say, for example, that they were told to be a doctor since that's what the entire family did, or many other things. But it doesn't matter what limiting beliefs or excuses you have: until you do what you love, your reality will be adversely affected—in relation to your finances, your health, or your relationships. Until you ride the resonating frequency of happiness, you will always encounter dissonance. It's up to you whether you wait another forty years or stop and get back on your path now.

This happened to me not too long ago. I was ignoring my reality, and my emotions eventually shut me down until I faced my subconscious yearning. I was working as a manager of a gym, and I had recently completed my personal training certificate. My theory was that I liked sports and I liked helping people, so therefore I must want to work in the fitness industry. After I arrived at my new job, the first couple of weeks—the honeymoon period—were fine, but I had this tiny feeling that I didn't enjoy the job. My justification (and most people will try to justify their negative feelings, which is a form of ignoring who you are) was that I had just got back from a five-year working holiday overseas and that it would be hard to readjust to the nine-to-five working world. So I kept on plugging along while my feeling of inner conflict kept getting worse and worse.

I am a very lucky person, because my body soon tells me that I am not following my path or that I am doing things that go against my natural energy vibration. So after a couple of months, I started getting depressed and unable to work, but I again ignored the signs until my body wouldn't let me suppress who I was anymore, and I started crying. I know that sounds funny, but every time I had to face the car to go to work, I started crying and was unable to stop. So I drove almost all the way to work, and then I drove to a park and tried to focus on the positives. I stopped my tears, but as soon as I thought I had myself together and started driving, I started to cry again. I didn't know what was going on. I completely broke down and had to call my father to pick me up.

I had to take a few days off work to listen to myself. I couldn't consciously work it out myself. I had reached the limits of my own thinking, so I went to see someone, a hypnotherapist. I realized that my subconscious had understood that this job was not for me, but my conscious mind was arguing that this was the only way to achieve my life goal of retiring at thirty-five. Therefore, if I couldn't do this job, I would have to give up myself—who I believed I was. That was my internal rationalization. No wonder I was upset!

I ended up realizing that it doesn't matter if you don't know how you will get to your ultimate dream goal. As long as you pay attention to your reality, you will know whether you are on track or not. If you follow the things that bring you happiness, then they will lead you to your reality. On the other hand, if you find yourself experiencing negative emotions, then be sure to change whatever brings you to that negative state.

So ask yourself what in your life could be causing dissonance at this moment. Is it physical, or is it mental? Is your body showing signs of illness that actually provide a clue or answer to whatever mental blockages you have been holding onto? It is always interesting to find these things out.

A good way to find out if there is dissonance in your life is to ask yourself the following questions:

1. Where do I want to be in life? What is my ultimate goal?
2. Am I anywhere near that reality?
3. What do I feel emotionally when I think about the journey to achieving my ultimate goals?

In the last question, if you are truly being honest with yourself, if you have *any* sort of negative emotions, that means that there is something to work on there—whether it's fear, a limiting belief, or a sense of being overwhelmed. Something needs to be to unlocked, addressed, and eliminated to help you evolve to your UFH.

There are generally four stumbling blocks that keep us from moving toward achieving our goals:

1. fear
2. limiting beliefs
3. excuses
4. frequency change/goal change

Fear

Fears are beliefs that have been so concretized into the foundation of your being that they become a reality for you. The good news is that they are ultimately still beliefs, and beliefs can be changed. As I mentioned before, in the words of Esther Hicks speaking as Abraham, "a belief is only a thought that you keep thinking." So if you want to change a belief, start thinking the opposite. Fear is a lot more deeply embedded than just in the mind. It will produce emotional attachments, and it will be so deeply rooted in your personal *path of affect* that it may be a part of your reality. A fear of failure might show up as a tendency to resonate with ridiculing people for making mistakes or tuning into stories on TV about people who become complete failures in life. The problem with fears is that they become so embedded in your *path of affect* that your reality reflects them. So the first step is to become aware of the paradigm and of the fact that you have an unrealistic fear.

Some fears are okay, since their purpose is to prevent you from jumping off a cliff or diving into a piranha tank. Fear originated to help us survive. However, fear of failure or fear of success can be completely debilitating and prevent you from ever achieving anything in life. So once you realize that you have an unreasonable fear, you can use some of the resources in the next section to help eliminate it and evolve to a higher frequency so that your reality resonates with nothing but success.

Limiting Beliefs

> *The power of belief can change your reality, so if you don't like your reality, change your beliefs.*

This is very powerful. But what if you don't realize that your beliefs are limiting you? This really happens. I have met a lot of people who are so limited by themselves that they don't realize it. Just listen to their language: "I can't do that—I'm not good at math." "I can't succeed in business, because I never went to school." "I am no athlete—I can't catch a ball to save my life." All these are limiting beliefs people have given themselves. Some will never really affect them, because their UFH is not in that area, so they aren't too concerned. But if you absolutely love doing something but can't do it for some reason, then stop and analyze what that reason is. Why are you really stopping yourself? Is it a fear or a limiting belief?

Let me give you an example of a limiting belief that stopped me from doing something for years. I had an incident with a treadmill one year. It had a computer malfunction and completely threw me off the machine. I wasn't injured too badly, but the shock and fright of it installed a limiting belief in me. I believed that I couldn't run above 9 km/hr, because that was the speed at which it threw me off. I was fine running up to that, but if I tried to go higher, I would freak out and jump off the machine. It completely ruined treadmills for me. So I gave up on them. Stopped altogether. Then a few years later, I had forgotten about what had happened. It was early one morning, and I got on the treadmill half asleep, listening to my dance music, off in my own little world ... and then I realized I was running at 12 km/hr. I couldn't believe it! My sleepy self hadn't woken up my limiting belief yet, so I had inadvertently defied it. It made me realize the restriction I had put on myself. Needless to say, I told that belief never to come back—it was no longer needed.

This is a light-hearted example. However, there are more serious limiting beliefs that could be stopping you from reaching your goals. I told you of the one I had created about guys—that my father was nice and cheated and that if nice men cheated too, how could I ever know for sure that a nice guy wouldn't cheat on me? I realized that I was the only person in this equation that could control that—it was all in my head. I worked through releasing this limiting belief with some exercises by Byron Katie (I'll talk about her in the resources section) and managed to eliminate it. That was an important step, because that limiting belief was stopping me from ever trusting men at all and completely ruining for me the possibility of ever having a loving relationship.

There are many different limiting beliefs. Listen to your language and look around at your reality—see what is showing up or not showing up because of whatever limitations you are putting on yourself. The next section will provide many different ways of ripping through these crazy beliefs to help you evolve and reach your pure, undisturbed frequency of happiness!

Excuses

About 99 percent of excuses are never really valid. They are the layers of bullshit we place around a fear or limiting belief so that we don't have to face the real truth. "I can't go the gym because it's too busy." "I can't go after work because it will just wake me up, and I won't sleep." "I can't achieve my goals because I am too busy." Any excuse—and there are an infinite number of them—will generally be hiding something more. The real truth of things remains beneath the surface as long as you keep playing the game of bullshit.

Now I may sound a bit harsh, and there are definitely occasional unfortunate incidents that can be valid excuses, like natural disasters, family emergencies, or the physical impossibility of getting to the only place where you can perform your activity. But short of something like that, there is always a way to do something.

So if you know someone who really wants to write a book (this *may* be an example of one of my excuses) and you ask them why they aren't going to the library like they said they would, and they reply, "Can't be bothered," then I dare you to delve a little deeper. Say, "Really, you can't be bothered to do the thing that makes you happy?" "Yep, I am just too tired …" "Really? but you are going to stay up late and watch TV. Why are you really not doing it?" "I have to make dinner, and I'll get hungry, and then I won't be able to think." "You managed to forget to eat the other day when you wrote. Why is this different?" "Okay, I'm afraid that if I finish and people don't like it, I will have failed."

Okay, so it might not be as easy as that to get to the bottom of your procrastination or excuses, but if you find yourself making excuses, that is the first indication that there is something more bothering you. I actually had that fear for a while, and then I went to my kinesiologist, who helped me release that thought and energy pattern, and amazingly, that day I wrote pages and pages!

Excuses are just layers of bullshit covering up a fear or limiting belief. If you find yourself using them, you need to face the music and move to evolve past them. The more limiting beliefs and fears you crush, the closer you will come to having your reality constantly mirror your UHF.

Changing Frequency

This can be a tricky one, because no one likes a moving goal. If soccer were played so that every time you got close to the goal, it moved somewhere else, I think everyone would give up pretty quickly—especially if you never knew where the goal was going to move.

This will happen on your journey to reaching your UFH. I've constantly had to evolve and change—but that's the journey. It's all part of finding that frequency of happiness. You may think you have it, and at first, it seems like bliss. Then all of a sudden there is a small dissonance in your happiness, and this eventually spreads until you realize that this isn't quite what you wanted to do. The idea is to keep searching for that resonance. If it shifts, you need to follow it. I actually had to keep following frequencies that shifted and then shifted again. For example, I loved learning about the power of the mind, so I decided to study life coaching. Once I achieved that, I thought I would be at my frequency, but it shifted again. I started building a life coaching program and completed the proposal, and then the frequency shifted again. I started studying consciousness, and eventually it was this book that satisfied my frequency needs. However, I don't think I could have written this book had I not gone through the other frequencies. As long as you follow your instincts and what makes you happy at the time, it will take you to where you are meant to be.

This shift in frequency is an evolution, and as long as the shift always follows the frequency of happiness, you are evolving in the right direction. You may have experienced this growing up. I know I did. Going back to the radio analogy, just think about the radio stations you've liked in your time. Personally, as a child I liked to listen to any radio station or music that played kids' songs, because they were fun, which was the frequency that resonated with me at the time.

As a teenager, I preferred the alternative stations that played angry, depressed music, because they resonated with my personal frequency at the time of not knowing who I was, my parents not understanding me, and my generally rebellious mood. Through my twenties I preferred the frequency of R&B, dance music, and the top forty—the frequency of partying and living life. Now, my frequency has settled somewhat. I still listen to some of the music I loved growing up, as those frequencies are still part of me, but they don't dominate who I am. Now, I like to listen to positive radio stations, those that help empower people. I like the music and songs that make me a better person for listening to them—not necessarily in one particular genre but across the board. And my frequency isn't affected by what others think is cool anymore—hence the stability in my frequency. I am happy with who I am and what I want to listen to without being concerned about what other people think.

Our frequencies change, and as they change, we match different people. The people I hung out with during my teenage years, who matched my frequency and resonated with me at the time, no longer resonate with me. Not only have I changed, but they've changed their frequencies too. Some stayed at that frequency level and will be there for the rest of their lives, but we grew apart—in other words, our frequencies no longer matched. The radio station I was tuned to couldn't quite receive the frequency they were on.

It is hard at those times when you feel that you're changing and your friends don't seem to be interested in the things you talk about or the direction you're taking. They can remain friends, but you feel like you aren't as close anymore. People often stop themselves from searching for their true resonance because it means that they will no longer resonate with longtime friends. Unfortunately this can happen a lot, and you are sacrificing who you truly should be! True friends will still be friends; they just might not understand the particular frequency you take. You can tune into your friends' radio stations to have a listen and hang out, and as long as you always go back to your true resonating frequency, you will be happy.

It's important not to get caught up in your environment and the people around you, especially if they aren't resonating with your higher frequencies. Understand that you will start tuning into people in your reality that resonate with your UFH. Peirce warns us of one issue, though: "There's a problem, however, and that is due to your body's tendency to resonate, like a tuning fork, and to change its vibration to match the frequencies of your environment."[166] You may start tuning into people that aren't on your UFH if you aren't able to really get into your UFH on your own. It is easy to be influenced by others and then tune into that environment. Just look at the people who are originally quite good but get sent to jail for some reason, and it makes them worse because of the frequency they end up tuning into and eventually resonating with. If you aren't strong enough to resist this energy shift, then that will very likely happen. You will know that it isn't really for you, though, because although you tune into these frequencies, you can still feel a deep dissonance with them. That is your intuition telling you that you haven't found your UFH.

Stay aware and responsible to your path. If you know you are vulnerable to other frequencies, surround yourself with the type of people who resonate at your UFH. That is the best way to help you to start resonating at that frequency, since you will be surrounded by it. Follow your path and stay in tune with your internal resonance radar. Resonance is your GPS, the UFH is your destination, you are the driver, and your thoughts are your steering wheel. It is up to you to use your thoughts to drive you in the direction of your UFH, using resonance to help you along the way.

RESOURCE OPTIONS—THE ULTIMATE FREQUENCY TOOL KIT

Whether you're trying to work out your ultimate frequency, which limitations or fears you need to overcome, what strategies you need to achieve your reality, or simply how to maintain your UFH, this section will provide you with all the resources you need to help you achieve your ultimate reality. I believe that there are three phases you will go through, sometimes at the same time. Generally, you will go back and forth between all of them. These phases are

- evolution,
- problem solving and creating strategies, and
- setting your frequency dial.

The first two will be needed as you move through your journey. The last is something that you should do every day in one way or another.

The following resources are ones I have used, come across through my research, or discovered along my own path or from other people's recommendations or success stories. It is up to you to pick the resources that resonate with you at the time. One thing may be far from connecting with you, but something else may just be the right thing at the right time. Later on in your journey, be sure not to dismiss things that you didn't try previously, since they might be right for you at another time. The resources listed below may lead you to find others that help you, or they may even lead you to discover a unique blend that is the key to your evolution and to attaining your UFH.

EVOLUTION

Every single one of us has weaknesses, limitations, and beliefs that hold us back from achieving a life of infinite happiness. We can get distracted by what we see rather than what is inside. We can forget about this new paradigm completely and return to believing that what we see is what we get. We are consciousnesses having a human experience to learn to overcome some subconscious issues. Whenever you come across conflict, it is a direct reflection of something you have yet to learn about yourself. Whether it is a fear, a limiting belief, or some issue that you need to get over from a past relationship in this life or a previous life, there is always something to keep us evolving. The more limitations and fears we can release, the closer we come to reaching a permanent level of UFH. To help do this, we sometimes have to delve into our thought patterns, our psychological backgrounds, or our habits or even look at what our physical ailments are telling us we have been holding inside for too long. As you read earlier, I held onto my limiting beliefs about men for ten years. This ended up affecting my physical being and forced me to have some of my cervix cut out. Your body is an indication of what is going on at the energetic, emotional, or belief level. The following resources are there to help you overcome and break through these barriers, blockages, and weights that are holding you back from moving toward happiness. This certainly isn't an exhaustive list; however, it will definitely start you on your path.

Neurolinguistic Programming

NLP, or neurolinguistic programming, helps you to reprogram thoughts that prevent you from achieving your goals and dreams. NLP should always be used to help people achieve their goals and dreams and never for anything negative. The good thing about NLP is that the person who uses the NLP tools has to want to use them for them to actually work. You need to be ready to let go of your limiting beliefs and fears and want to move forward. One of the greatest NLP trainers is Chris Howard, who developed the Academy of Wealth and Achievement (http://www.beyondnlp.tv/register). Their programs teach and train in NLP, and you'll find that it's a brilliant tool to have. Otherwise, you can always look up other trained NLP practitioners to help you break through your destructive thought patterns.

Life Coaching or Psychology

The main difference between life coaching and psychology is that life coaching works on moving forward from where you are, whereas psychology or counseling often work on pinpointing the issue in the past and working through it. Both have their advantages and often work well in combination, releasing old past wounds and moving forward with new strategies. Find coaches and counselors who are recommended by friends or family. Remember, you might go through several before you find one that resonates with you. You can often find life coaches that specialize in NLP, so that is also handy.

The Power of Now by Eckhart Tolle

The Power of Now by Eckhart Tolle is a brilliant book. Essentially, it talks about being 100 percent focused on the present moment—in other words, not thinking about what someone said yesterday and not worrying about what to wear tomorrow. If you are fully present in the moment of now, you will be truly at peace. When you are truly at peace, you are able to feel the real you, not the ego-you that is worried about what everyone else thinks. If you feel the real you, it will help you to hear what frequency to follow. If this sounds like a method that might suit you better, read his book—it is life changing.

Byron Katie's Process, The Work

Byron Katie wrote the book *Loving What Is* and is famously known for her revolutionary process called *The Work*. Essentially, she asks four questions that help you to adjust your beliefs to what is, not what isn't. If you have beliefs that just can't work, you will always be going against the energy flow, so her very simple method of four questions can help you adjust your beliefs to go with the flow of your energy. There are excerpts from the book and free worksheets on her website, www.thework.com.

Tapping or Emotional Freedom Techniques

"Meridian tapping combines ancient Chinese medicine with modern psychology to create a remarkable new approach to stress."[167] By using a method that combines tapping the recommended points on your body while repeating specific personal phrases out loud, you can learn to help yourself release the energy and thoughts that you have been holding onto. I watched a DVD on tapping and found it remarkable. A great book on it is *Discover the Power of Meridian Tapping* by Patricia Carrington. This method is great not only for breaking through some limiting beliefs but for getting in touch with the energetic level as well.

Pranic Healing, Kinesiology, Acupuncture, Aura Reading, and Other Energy Healing Forms

Pranic Healing works with chakras and *prana* (life-force energy or the ocean of energy).

Kinesiology works with your internal personal energy structure, which can communicate what is wrong with you physically or emotionally. It's very interesting, and if you find a good kinesiologist, he or she will pick up on what is lacking in your physical body and what it needs.

Acupuncture is the Chinese method of unblocking any energy blockages you may have. I find it to be good for healing at the energetic level; however, if you don't fix the thoughts and beliefs that are causing the blockage in the first place, they will keep affecting the energy.

Aura Reading is a process in which a practitioner is able to see, feel, or sense your aura, including any blockages or disruptions. A good aura reader may even be able to get information from the ocean of vibrating holographic information surrounding you, as in my case. Again, it is helpful to get the information and then utilize the other resources to help you break down the thought patterns that created the disruptions in the first place.

All of these types of healers are fantastic at working at the physical level as well as the energetic level. Often, they will even reach down through everything else to your thoughts. Those that work right into the *path of affect* will help to destroy any negativity that is holding you back. By reading your energy, they will often be able to pick up on physical symptoms or emotional stresses. It can be beneficial to combine what you learn here with other methods, especially once you learn what thoughts and limiting beliefs are affecting your energy or physical body.

You Can Heal Your Life and *The Secret Language of Your Body*

These two books by Louise Hay and Inna Segal are both about looking at the specific ailments in your body and analyzing what mental or emotional issue is linked to them. From there, they are about healing yourself at that level. They both provide suggestions and methods to overcome the beliefs or limitations you have put on yourself. These books are great to have around the house and ready for use in case of illness, because they can help you to cut it off before it gains any strength or has any serious effect on you.

SOLVING PROBLEMS AND CREATING STRATEGIES

There will be many times throughout your journey when you stray off the path, and you may not know what you want to do. Friends' ideas aren't working, and you just feel stuck. You either can't overcome a problem or you don't know where to go. You don't feel that it's really enough of a need to justify paying for a life coach; however, it is significant enough to keep you from moving forward.

I would, first of all, always recommend a life coach since, as with a personal fitness coach, you will push yourself further and achieve more with someone coaching you regularly on the side. However, I know that there are times when you just want to try to work it out yourself. Well, do I have an amazing little handy hint!

This method for working out answers to problems or generating ideas for moving forward is known as the GROW method, and it was first developed by Graham Alexander, Alan Fine, and Sir John Whitmore.[168] At the Life Coaching Academy where I obtained my certificate, we expanded it to IGROWN. Each letter stands for a step in the process of unraveling your problem, finding a solution, and moving forward:

Issue
Goal
Reality
Options
Way Forward
Nail Down

Normally it's better if someone you don't know does this with you, since they can't lead you down any particular path or make any previous judgments. The best possible way is to find a life coach yourself. However, if you aren't ready for that and want to give it a go yourself, this is what you should do. Find yourself a place away from distractions, turn off your phone, and make sure you won't be interrupted. Grab some paper and a pen, and follow these steps:

1. ISSUE

Write down what issue you want to work on. It could be that you want to work out what your frequency is, what job you should do, or what makes you happy. If it helps to just start writing about how you're feeling, then start with that. Once you have established a general issue you want to work with, move to the next step.

2. GOAL

This is where you need to put your issue into a one-sentence goal that describes what you would like to achieve from your session. You also need to put it into a short, concise sentence in the positive. So put "I want to find out my true inner frequency" rather than "I don't want to be unhappy anymore." Put "I want to find a job that will make me happy" rather than "I don't want to do this job anymore." Once you are confident you have a goal and it is positive, move to the next step.

3. REALITY

This is where you have to explore the current reality that is showing you that you have an issue or problem to work on.

So write down the answers to some questions:

- What is missing from your reality that you would like to have?
- What is happening now that is good and that can help you?
- What have you done so far to improve things?
- What were the results that you got from doing these things?
- What hurdles are in your way that are stopping you?
- Is your original goal still what you're looking for, or has it changed while you were working out your reality?

4. OPTIONS

This is the really fun part! This is where you get to explore any options you have to help you achieve your goal. There are absolutely no restrictions here—no idea is a bad idea, since you may get something good from something you thought was impossible or ridiculous. Take ten minutes to sit and just write down all the ideas you can about how to achieve your goal. List as many as you can—try for at least twenty. Force ideas out of your head.

Here are some questions that might help you get some more ideas:

- Think of someone you idealize who has achieved your goal. Imagine you went up to this person and asked him or her what you should do. What would the person say?

- If you are stuck and not sure what to answer, then a great question to ask is, "If I knew the answer, what would it be?"
- What would you do if money weren't a problem?
- What would you do if the hurdle(s) you identified in step 3 weren't an issue?

Now go back over all your scribbles and decide which one stands out to you. It might feel like it has a bit more energy around it. Close your eyes and feel what stands out to you. Once you pick one option, I want you to first check whether it will help you move forward toward your goal. If not, pick another until you find the right one.

5. WAY FORWARD

This is where you take your idea and start putting it into practice and locking it in.

- Choose a target date for your goal.
- Write down the steps you need to take to achieve your goal.
- Do you need to get anyone else involved in helping you? What will they do? When will you tell them?
- What is your first step? When will you do it?
- Can you think of anything that might prevent you from achieving your first step? If so, how likely is it to stop you? If it is likely, what can you do to prevent it from holding you back?
- Is your target date still possible?
- Is there anything else that you need to consider?

6. NAIL DOWN

Now this is where you nail yourself down to these steps. And you have to be honest with yourself; otherwise you are only doing yourself harm. Answer the following questions using on a scale of one to ten:

- How strong is your intention to take the first step?
- How high is your enthusiasm for taking the first step?
- How strong is your commitment to taking the first step?

Now if any of those steps scored a seven or below, you need to find out what has to happen for you to raise that score. If that has to be your new first step, then do that.

This coaching method can really be used for any issue—you just have to very honest with yourself. If you aren't, it would be best to find a life coach that can assist you in finding your own answers.

SETTING YOUR DIAL TO THE ULTIMATE FREQUENCY OF HAPPINESS

This is an activity that should be done regularly—at least daily. It is about tuning into your frequency of happiness so that you constantly are in tune with it so that it can resonate in your reality. This can be done from all levels on the *path of affect*—from your thoughts and beliefs all the way through to your emotions and energy. Even when creating a new reality, it's helpful to know that shifting your inner frequency can trick you into it. All these tasks are about connecting to the frequency and thoughts that make you happy, that make you love and be grateful for what you have. Once you fully—right down to that Planck scale—really feel and believe in the frequency you want to achieve, then your reality will mirror it. As I said, it's not the law of attraction, but reality through resonance. If you want a reality full of happiness and love, you need to resonate at that frequency. To help you do this, here are some resources you can use:

Affirmations Affirmations are short sentences of a positive nature written in the present tense and aimed at helping you reprogram your beliefs. Gerber cites Emile Coué as one of the early pioneers at using them for healing: "The use of structural affirmations in healing goes back nearly a hundred years. Emile Coué, a French pharmacist around the turn of the century, opened a free clinic. Coué's work on autosuggestion stemmed from his belief that people's thoughts eventually materialized as physical reality. As such, he believed that dysfunctional thinking patterns could actually worsen and perpetuate different forms of illness."[169]

Although they may sound silly at first, if you are disciplined with them, they really can start to change your attention and focus it onto new realities and beliefs. Coué's affirmation is one of the most famous: "As a basic summary of early-level programming affirmation-making, start by writing out a simple script to follow for doing daily affirmations. One of the easiest affirmations is Emile Coué's simple phrase 'Every day in every way I am getting better and better.' The phrase can be modified to say … ['healthier' or 'wealthier'] as well as other variations on a theme."[170]

Gerber goes on to say, "Affirmations and visualization techniques help to reprogram your consciousness and belief systems to expect abundance and perfect health in life as opposed to operating from an attitude of fear, frustration, stress, and the paranoid expectation that everyone in the world is out to get you or to give you a hard time."[171] They basically plant the seeds right from the beginning of the *path of affect* into your thoughts and beliefs, which, as you know, will affect your emotions, your energy, and ultimately your reality.

Frequency Shifters These are similar to vision boards; however, I have pimped things up a bit. Vision boards were about sticking pictures of your ultimate reality on a board in your room to help attract your reality. Frequency shifters are similar, but they are truly about shifting your frequency to the level of that UFH. So it is not only about sticking pictures of your ultimate reality on the wall—it's also about adding short-, medium-, and long-term goals in the form of affirmations. As each one is achieved, you put a star or checkmark next to it so you can see the frequency shifter coming to life. The more you see it coming to life, the more real and frequency-shifting it becomes. Generally you have a frequency shifter for each ultimate frequency in each area of your life—health, career, wealth, spirituality—or you can even have them as goals.

How to make one:

1. Divide your cardboard into thirds. In the top third, write an affirmation of your ultimate reality. (For example, I have four different frequency shifters on my walls, each with an affirmation as its title: I am a fit, healthy, and a toned size 10; I wrote a number-one international best-selling book; I travel when, where, with whom, and how I want; I am financially free with over $1,000,000 a year in passive income.)

2. Under the affirmation title, paste your photos and pictures of how you visualize the affirmation. (I have pictures of flat-toned stomachs and healthy, energetic women under my health frequency shifter and pictures of publishers, the New York Times bestseller list, and so on, under the frequency shifter dedicated to my book.)

3. Divide the rest of the cardboard into three for short-, medium-, and long-term goals. Under these sections, again in affirmation format, write the relevant goals. I wrote them on colored paper and stuck them underneath each heading with a big square ready for me to stick a star on it when I achieved it.

4. Once it is done, stick it somewhere you can see it every day.

5. Make a conscious effort to connect with each frequency shifter every day, even if it is for only a minute. Really feel the presence of that frequency shift as being real.

6. Watch your reality resonate your frequency shifters!

Mind Movies I came across this on the internet at www.mindmovies.com. They are brilliant and are pretty much live video versions of frequency shifters. They take the pictures and put music to them with the affirmations. You can create your own and even have them playing in the background of your computer. This is a brilliant way to subconsciously install the frequency of your reality.

Music This is something basic you can do. If you are feeling low and want to connect with your UFH, find a song that really picks you up, takes you to another level, and makes you happy. Just by doing this, you will shift the reality around you. It is really a great way to get yourself out of a rut. It may feel stupid at first, since you won't be at that frequency, but by the end of the song, you will really have raised your vibration from the subatomic level right through to your reality.

Meditation For thousands of years, monks have used meditation to achieve happiness, to become fully present in the now, and to connect with the ocean of energy, becoming totally in tune with its vibration. If you can connect with the ocean of energy, you will feel the UFH. The natural resonance of the ocean of energy is the Ultimate Frequency of Happiness. Attain that, and you will have found a great way to resonate with your reality.

Tuning Forks and Singing Bowls There are actual tuning forks that you can buy that are for different emotional frequencies. Ancient Eastern cultures and religions like Buddhism use singing bowls that, when struck, send out a meditational frequency to help you relax and meditate. I recently experienced a device that has been created using two quartz crystals hooked up to a computer. You hold the quartz crystals, and it sends a certain frequency for happiness or relaxation through the quartz to help shift your energy frequency. It is really quite fascinating!

These methods are all about feeling the frequency. Although you need to be able to overcome problems and work on eliminating the negatives that hold you back—which is what the other two phases are about—it's this phase that you need to connect with on a regular basis. It's really about tuning into your internal frequency—shifting the dial so that you tune into everything that resonates with the frequency you want. If you start your day like this every day for thirty days, you will be amazed at how quickly your life can shift! It is totally up to you to decide to take on this challenge. Experiment with this paradigm and give it a go! What do you have to lose? The worst that can happen is that you that feel a little happiness, and if that doesn't suit you, you can always go back to letting the external world be in charge of your destiny.

THE SEARCH FOR RESONANCE

You are on your way to a new paradigm. You have your ultimate frequency tool kit all stocked up, and you are ready to experience the UFH. I want to leave you with one last thing: a small guideline to help you stay aware and keep your tuning fork tuned to your ultimate frequency every day.

YOUR REALITY IS YOUR RESPONSIBILITY—YOU HAVE TO BELIEVE IT TO SEE IT

- ***Your consciousness is the driver—that is to say, you are.***
- ***Thoughts and beliefs are your steering wheel.***
- ***Resonance is your GPS (look closely at what resonates in your reality to tell you if you are on the right path).***
- ***The Ultimate Frequency of Happiness is your destination.***
- ***The ultimate frequency tool kit is there for first aid and roadside assistance.***

<u>Do three things daily:</u>

1. ***Feel your ultimate frequency.***
2. ***Reality check—if it's not working, what is really going on inside? Open your tool kit.***
3. ***Let go and have faith that you are always on the right frequency.***

A famous line from the poem "Invictus" by William Ernest Henley goes like this: "You are the master of your fate, you are the captain of your soul." You are the center of your own reality, and you are the only person to blame if things aren't going according to plan. But you can choose what beliefs to follow and what thoughts to think, which in turn will create the reality that is your life. A quote from the Jackie Chan movie *Karate Kid* echoes what this book is about: "If you think with your eyes, you are easy to fool." Either start looking within and changing your reality from the inside out, or keep looking outward and letting your surroundings control your fate.

Chapter 5: From Planck Size to Beyond Dimensions

Planck → Individual Views

"Your reality is your own responsibility. You have to believe it to see it!"

At the most basic level of life is the Planck scale. In this subatomic world, the possibilities are limited only by the imaginations of the consciousnesses that influence them. As conscious beings, it is our responsibility to recognize the power of our influence and be mindful of its consequences—to be aware of the thoughts and beliefs to which we constrain ourselves and others. Whether or not this paradigm has fully sunk into your being yet, it is difficult for you to deny the substantial amount of scientific evidence that is pointing toward the idea that the responsibility is ours. We are the center of our own realities, and it's up to us to take on this challenge with positivity and love.

With your conscious awareness now awakened, your mission is to foster the beliefs that support you and raise you to your UFH. Let go of limiting and destructive beliefs to make room for new radiant, attractive, fun, and happy beliefs. The ego is powerless and no longer able to hurt those around you—especially you. Having the freedom to choose raises your frequency to a higher awareness—even if only on such a minute scale as the Planck scale—and the people around you notice a change. They notice a new you, a you that isn't limited by useless thoughts or by what other people think. They see a free spirit.

We are now in a new age in which conscious life is immortal. We never lose anyone—we are all part of each other, and the more love we have for ourselves, the more we love each other. We are all interconnected in a universal togetherness, and our mission is to help fully recognize our own potential and to help others do the same. Animals are able to help us with their expanded ability to sense frequencies and their ability to sense and bring our awareness to the diseases in our bodies. We no longer need to be influenced by the medical industry, since it is our thoughts and beliefs that heal us and that are the most powerful drug around. Religion has a chance to be more popular than ever, with each individual embracing whatever religion is right for them.

As individuals, it is our responsibility to accept this new universal paradigm—to accept everyone as they are—for now we can recognize that their reality is right for them as ours is for us. No longer is there an ego battle over who is right and who is wrong. We are all right within the frameworks of our own realities. No one can say otherwise. With the battle of right and wrong eliminated from existence, peace is attained. A love and respect for everyone as they are is an ever-present possibility in this positive, energy-based reality that we live in and of which we are a part.

The following is a set of guidelines designed to help you lead a universalistic life:

UNIVERSALISTIC GUIDELINES

1. Respect each other's realities—everyone is right.
2. Find your UFH purely through thought and without the help of external influences.
3. Surround yourself with others that support the reality you wish to live in.
4. Be sure that all your relationships are about exchanging energy—a constant giving and receiving of energy with the other consciousnesses. Build and support each other to be the best that you can be.
5. Never take energy from others or restrict their realities. If you find yourself doing this, stop and listen to the thoughts that led you to this place.
6. Ultimately, your goal is to create a universalistic reality that resonates with your UFH.

Individual → Global Consciousness

Now—Mass Media

Unfortunately, the present global consciousness is dependent on the physical reality we see around us, and the media has the responsibility of pointing out what is going on outside our immediate reality. Every day, events are brought to our attention that show our backward way of thinking, which, if left unchecked, will spiral us downward to inevitable self-destruction. Glimpses of this can be recognized in events and issues like

9/11, climate change, and now the latest—the global financial crisis. It is backward because of our current paradigm—you have to see it to believe it. The media point out everything they see for us to believe. Where this goes wrong is in the pure motivation of some segments of the media to feed on fear, because it is fear that sells magazines or has people watching television. So in the last few decades, with the global consciousness becoming more connected by internet and multimedia, the media has been able to capture more and more people, making them dependent on external realities for their cues about living. They will take any story and turn it into a fear frenzy—you only have to watch certain current affairs shows to see this syndrome at its worst: "The Cream That Will Make You Eliminate Your Fat," "The Bank that Stole Your Money," "The Supermarket That's Poisoning Its Customers"—anything to get people to watch. When they can, they combine their forces to drag out horrific events like 9/11, climate change, and the global financial crisis for several years, reminding us constantly of our mortality.

Well, with this new paradigm, times are changing. And if I can be so bold, I'd like to point out something that the media might possibly learn from the children's movie *Monsters, Inc.* In the beginning, the monsters run their city's energy purely from the fear of children, since it is the only way they know. For generations, this has been what worked, so they decided to keep the blinders on and keep doing it that way. They soon find out, though, that more energy is produced from children's laughter and happiness than from fear. They can run their city with more than enough energy to go around purely on the frequency of happiness. This could be the story of our global reality as well. The frequency of laughter and happiness is much more contagious, positive, and energy-building than that of fear. We could move forward instead of backward. My hope is that a network or channel will start to break the mold and only highlight the news of the day that brings hope and that consistently builds on the frequency of happiness.

Oh, I know—I can already hear people protesting, "But you need to know what's going on in the world!" Why, if your reality is your own? If what you decide is true can affect your reality, why do you need to distract yourself and influence your emotions with a war that is happening on the other side of the world? Is that going to help you create the reality you want—if your thoughts and emotions are on war? Now here's a thought: What if you decide that your reality should be about peace and love? What if you decide to surround yourself only with things that help you to vibrate at the peace frequency? Won't your observations of the world as peace and love start to resonate in your reality? If you truly take on the frequency of pure peace and happiness, and if others around you then take on the same frequency—won't enough realities soon reflect this around the globe? The possibility of taking this leap is there. Are you game?

Future—Global Consciousness Program

A prayer in the state of Washington illustrates the effect of a combined global consciousness: "As a result of prayer, an ongoing prayer initiative was started to have 24/7 prayer in each county of WA state. Since we started crime has been a target and this year we saw crime in every major city decrease with Seattle reporting a 40-year low."[172]

Another experiment that illustrates the effects of mass energy focus was done by Dr. Masaru Emoto in Japan. He showed what can happen when we use the combined energy we can create as a group:

> In September 1999 ... 350 people had gathered on the banks of Lake Biwa, Japan's largest lake. I had gathered the group together in an attempt to clean the lake ... this large crowd joined forces in an affirmation for world peace that brought our voices and hearts together. Our chants could be heard around the entire lake, and there was a special feeling that made our spines tingle [energy]. Just a month after this event took place, a strange thing happened to Lake Biwa. The newspapers reported that the putrid algae that appeared each year and caused an unbearable stench had not appeared that year.[173]

Imagine if enough of us lived our lives around the created realities of a healthy frequency and included other countries that usually suffer from HIV in our reality as places of health and strength. Or what about shifting our realities to see peace in war-torn countries? These changes in our realities are possible from the Planck to the global scale.

A scientific study called the Global Consciousness Project (GCP) has also noticed that combined consciousness can affect matter, in this case, Random Number Generators (RNGs).[174] During the 9/11 attacks, there was such a significant reduction in randomness in the RNGs that showed that when the whole world focuses on one thing, matter can be affected. Imagine if this focus were deliberate and directed toward something more than just affecting hundreds of RNG machines. It is time to lift the lid on the possibilities. They are limitless, and our beliefs should reflect this if we want to take full advantage of this new paradigm.

If you are interested in these global-scale changes in reality, you may want to check out the following: Lynne McTaggart, author of *The Field*, is doing an experiment known as the Intention Experiment. You can read about it in her book of the same name[175] or go to her website (www.theintentionexperiment.com) and participate (or both).[176] The more we start to embrace these possibilities, the closer we will come to reaching a tipping point where they become more normal in everyday existence.

Universalistic View on Life

You only have to go online to realize how small and connected this globe is becoming. I was at a web conference the other day where Deepak Chopra had just come from a studio where he was presenting as a 3-D hologram to a group of people in New York (I was in Melbourne). Yet, though our globe is getting smaller, we are still holding onto the old ego battles of hundreds of years ago. Our minds need to catch up with the technology and science of today. We need to take a universalistic approach to life: "Only when man moves from his ego-centered self-image to a new image of universal human will the perennial problems that plague us be susceptible of resolution. Humanity must rise from man to mankind, from the personal to the transpersonal, from self-consciousness to cosmic consciousness."[177]

A universalistic view on life is about respecting every consciousness regardless of what it believes in as its reality. We are here to help others evolve as consciousnesses—that is, to help them be free of their limiting beliefs, free of their negative pasts and the wounds that hold them back. The more we evolve together as a global consciousness into a universalistic race and world, the more we raise our frequencies to new and higher ecstatic levels. Imagine walking around in a constant state of elation, meeting new and interesting consciousnesses. Imagine learning new and interesting things from them that we can then adapt to our realities if they resonate with us.

When the global consciousness begins to resonate with the Ultimate Frequency of Happiness, heaven will come to earth!

Global → Multidimensional

We live in a multidimensional universe. The old paradigm suggested that we look out into deep space to find our next adventure, while the new paradigm is showing us that it is right here, right now. It is time for us to embrace the possibilities of a multidimensional universe and see how this universe relates to our limited sense of dimension. Most ancient tribes and cultures around the world believe in a multidimensional existence with an afterlife and other beings that are a part of the universe. It seems highly improbable that the majority of tribes and cultures are wrong, especially now that science is also starting to point in the same direction.

If extraphysical consciousnesses inhabit the same space at the same time, how are they influencing us? Could they have anything to do with what is happening in our dimension? When things happen that we don't understand, rather than merely looking for the answer within our limited frequency range, we need to open our minds to the possibility of a multidimensional effect. What's going on beyond our senses? What can our multidimensional consciousness teach us? If OBEs can really enable us to cross over to other dimensions, then maybe we need to look at increasing our awareness of this and creating realities where OBEs are just another vehicle that we use in our everyday lives to learn and educate ourselves in this life and beyond.

If anything, I hope you walk away from this book with a fresh set of eyes that is able to see your potential and remove the restrictions you were placing on yourself through your limited views. Once upon a time, we thought the world was flat, and then it became a globe. Now we realize that it isn't confined to the physical—that it's a multidimensional universe of shifting frequencies and that we're at the center of influence in our own reality. What's next? Your reality depends entirely on where you want to take it!

May you resonate with your reality at the Ultimate Frequency of Happiness!

Afterword

I came to this theory through years of research. Each time I found new relevant information, my theories would slightly morph. One thing I have learned in my four years of research is that there is no such thing as absolute truth. An absolute truth is something that is absolutely the final answer—there is no other answer, that is it, and it will never change. Society started off with absolute truths but soon discovered that the more we learn and research, the less we know of absolute truth. Centuries ago, it was an absolute truth that the earth was flat. Until recently, Pluto was a planet. Now it's thought of as a dwarf planet. Realities are subjective. Think about this: I was always taught that the sky was blue, that water was blue, and so on. You were also taught that the sky was blue. But who's to say that the blue that you see isn't the same color as my green or pink? We will never know. Our labels are the same, but what we see through our eyes could actually be completely different.

The reason I am saying this is that the answer that I have currently researched and finalized is a snapshot of my current findings, which are a compilation of the current relative truths I have discovered. There could be a new discovery, or one of these truths could morph into a different truth, thereby revealing itself to be only a relative truth. In two, five, fifty, or even a hundred years, this theory might seem as true as the notion that the world is flat seems to us now. All I ask is that you absorb the information, take what you feel is true for you now, and research and grow with it. Find out what's true for you. Retain your capacity for discernment toward the ever-changing and evolving universe. Keep your eyes and mind open to what's around you.

At the end of the day, the reality you choose to follow will be the one you are able to verify. In the words of the International Academy of Consciousness, "Don't believe in anything ... Experiment! Have your own experience."[178] And that is something I want you to do. If you are at all intrigued by any of this information, then go out, find out more, and come to your own conclusions. You may discover some even more exciting phenomena.

Endnotes

1. Rupert Sheldrake, *Dogs That Know When Their Owners Are Coming Home and Other Unexplained Powers of Animals* (London: Hutchinson, 1999), 120–121.
2. Sheldrake, 24.
3. Sheldrake, 185.
4. Sheldrake, 199.
5. Sheldrake, 218–219.
6. "Journeys Out of the Body," YouTube video, 10:01, from a 1979 WPIX 5 San Francisco interview with Robert Monroe, uploaded by "MonroeInstitute," October 27, 2009, http://www.youtube.com/watch?v=wrLApcABHQw&feature=related.
7. Louise Hay, *You Can Heal Your Life* (Carlsbad, California: Hay House, 2004).
8. "About Anna," *Anna Meares*, 2010, www.annameares.com.au/about-anna.html.
9. Rhonda Byrne, *The Secret*, Web-based feature film directed by Drew Heriot (Prime Time Productions, 2006), DVD.
10. Louisa Gilder, *The Age of Entanglement: When Quantum Physics Was Reborn* (New York: Knopf, 2008), 6.
11. Stephen Hawking, Ph.D, and Leonard Mlodinow, , *The Grand Design: New Answers to the Ultimate Questions of Life* (London: Bantam Press, 2010).
12. Bruce Lipton, , *Biology of Belief: Unleashing the Power of Consciousness, Matter & Miracles* (Boulder: Sounds True, 2006), audio CD.
13. *What the Bleep!? Down the Rabbit Hole*, directed by William Arntz, Betsy Chasse, and Mark Vincent (Roadside Attractions, 2006), DVD.
14. Nonprofit organization that researches consciousness and human experience and potential with a scientific approach. For more information, visit www.noetic.org.
15. Lynne McTaggart, *The Field: The Quest for the Secret Force of the Universe* (London: Element, 2003) 27.
16. Hawking and Mlodinow, 131.

17. David Bohm, *Wholeness and the Implicate Order* (London: Routledge & Kegan Paul, 1980), 191.
18. McTaggart, xxvii.
19. Penney Peirce, *Frequency: The Power of Personal Vibration* (New York: Atria Books, 2009), 27–28.
20. Arntz, Chasse, and Vincent, *What the Bleep!? Down the Rabbit Hole.*
21. Bohm, 191.
22. Jill Bolte Taylor, , *My Stroke of Insight: A Brain Scientist's Personal Journey* (London: Hodder & Stoughton, 2008).
23. Jill Bolte Taylor, , "Jill Bolte Taylor's Stroke of Insight," video, *TED: Ideas Worth Spreading*, March 2008, http://www.ted.com/talks/jill_bolte_taylor_s_powerful_stroke_of_insight.html.
24. "The Electromagnetic Spectrum," illustration, *National Aeronautics and Space Administration*, last modified January 6, 2011, http://mynasadata.larc.nasa.gov/ElectroMag.html.
25. Richard Gerber, *Vibrational Medicine for the 21st Century: A Complete Guide to Energy Healing and Spiritual Transformation* (London: Judy Piatkus, 2000), 5.
26. Michio Kaku, , *Hyperspace: A Scientific Odyssey Through Parallel Universes, Time Warps, and the 10th Dimension* (New York. Oxford University Press, 1994), 153.
27. McTaggart, 101.
28. McTaggart, 108.
29. Peirce, 47.
30. Peirce, 47.
31. Kaku, 13.
32. Kaku, 132.
33. Hawking and Mlodinow, 115.
34. Hawking and Mlodinow, 116, 117.
35. Waldo Vieira, , *Projectiology: A Panorama of Experiences of the Consciousness Outside the Human Body* (Rio de Janeiro, Brazil: International Institute of Projectiology and Conscientiology, 2002).
36. Kaku, 117.
37. Gilder, 3.
38. Gilder, 6.

39. Dean Radin, , *Entangled Minds: Extrasensory Experiences in a Quantum Reality* (New York: Paraview Pocket Books, 2006), 14.
40. Bohm, 175.
41. Sheldrake, 164–166.
42. *The Quantum Activist,* directed by Ri Stewart and Renee Slade (Eugene, OR: Blue Dot Productions, 2009), DVD.
43. Radin, 14.
44. McTaggart, 223–224.
45. McTaggart, 224–225.
46. Arntz, Chasse, and Vincent, *What the Bleep!? Down the Rabbit Hole.*
47. Peirce, 28.
48. Arntz, Chasse, and Vincent, "Entanglement—the other side of the coin," interview with Jeffrey Satinover, , *What the Bleep!? Down the Rabbit Hole.*
49. Tony Rothman , and George Sudershan, , *Doubt and Certainty: The Celebrated Academy Debates on Science, Mysticism, Reality* (Cambridge, Massachusetts: Helix Books, 1998), 289.
50. Edgar Mitchell, DSc, , *The Way of the Explorer* (Franklin Lakes, NJ: New Page Books, 2008), 207.
51. Mitchell, *The Way of the Explorer,* 179.
52. Michael Talbot, *The Holographic Universe* (New York, NY: Harper Perennial, 1991), 54.
53. Barbara Ann Brennan, *Hands of Light: A Guide to Healing Through the Human Energy Field* (New York: Bantam Books, 1988), 27.
54. Hawking and Mlodinow, 68.
55. Arntz, Chasse, and Vincent, "Dr. Quantum—Double-Slit Experiment," *What the Bleep!? Down the Rabbit Hole,* accessed January 17, 2012, from http://www.youtube.com/watch?v=DfPeprQ7oGc
56. For a better understanding of why this happens, refer to the following: "Schrödinger's Quantum Mechanics Part 2," YouTube video, 11:12, uploaded by "bestdamntutoring," Jun 5, 2011, http://www.youtube.com/watch?v=ZYacDt0m3yY&feature=watch_response.
57. Radin, 224.
58. McTaggart, xx.

59. McTaggart, 151–152.
60. McTaggart, 161.
61. Amit Goswami, , *Physics of the Soul: The Quantum Book of Living, Dying, Reincarnation, and Immortality* (Charlottesville, VA: Hampton Roads Publishing Company, 2001), 264.
62. For more information, visit www.noetic.org.
63. For more information, visit www.iacworld.org.
64. Vieira, 34.
65. Vieira, 34.
66. "Holosoma," *International Academy of Consciousness,* November 8, 2010, http://iacblog-english.blogspot.com/2010/11/holosoma.html.
67. Goswami, 94–119.
68. Goswami, 104.
69. Richard Matheson, *What Dreams May Come,* feature film directed by Vincent Ward (Interscope Communications, 1998).
70. Hawking and Mlodinow, 46.
71. Esther and Jerry Hicks, 223.
72. Hawking and Mlodinow, 47.
73. Gregg Braden, *The Science of Miracles: The Quantum Language of Healing, Peace, Feeling, and Belief,* directed by Hay House (New York: Hay House, 2007).
74. Braden, *The Science of Miracles.*
75. Masaru Emoto, , *The Hidden Messages in Water* (New York: Beyond Words Publishing, 2001).
76. McTaggart, 175–176.
77. "Goals & Objectives," *International Academy of Consciousness,* 2010, http://www.iacworld.org/english/academy/goals-objectives.
78. Peirce, 168.
79. McTaggart, xxi.
80. Mitchell, *The Way of the Explorer,* 207.
81. Gilder, 14.
82. Robert Monroe, *Journeys Out of the Body* (New York: Doubleday, 1977), 265.
83. McTaggart, 166.
84. Peirce, 160.
85. Peirce, 160.
86. Peirce, 166.

87. Sheldrake, 236.
88. Mitchell, *The Way of the Explorer,* 175–176.
89. Sheldrake, 265–266.
90. Sheldrake, 258.
91. Sheldrake, 201.
92. Vieira, 956
93. Sheldrake, 218–219
94. Sheldrake, 189
95. Gerber, 23.
96. Gerber, 13.
97. Gerber, 13.
98. Goswami, 77.
99. Goswami, 77–78.
100. Goswami, 88.
101. Goswami, 88.
102. Marie D. Jones, *Modern Science and the Paranormal (Haunted: Ghosts and the Paranormal)* (New York: Rosen Publishing, 2009), 50.
103. Monroe, *Journeys Out of the Body,* 272.
104. Vieira, 605.
105. Jones, 53–54.
106. Brennan, 39.
107. Gerber, 26.
108. Jones, 54.
109. Goswami, 210.
110. Jones, 79.
111. Talbot, 142.
112. McTaggart, 185–212.
113. Jones, 87.
114. Jones, 84.
115. Radin, 248.
116. Talbot, 208.
117. Brennan, 39–40.
118. Brennan, 26.
119. Ted Andrews, *How to See and Read the Aura* (Saint Paul, MN: Llewellyn Publications, 1991) 3–4.
120. Andrews, 44.
121. Andrews, 45.
122. Gerber, 18–19.

123. Gerber, 51.
124. Myss's model of the chakras differs from more standard ones in that it is a synthesis of the ancient Hindu chakra model, the seven Christian sacraments, and the ten sephirot of the kabbalah as well as her own work as a medical intuitive. For more information about Myss's model, refer to her book *Anatomy of the Spirit: The Seven Stages of Power and Healing* (New York: Three Rivers Press, 1996). See also chakra in the glossary definitions.
125. Caroline Myss, *Energy Anatomy* (Boulder, CO: Sounds True, 1997), audio CD and booklet, 3.
126. Myss, 2.
127. Talbot, 230.
128. Talbot, 233.
129. Kenneth Ring, *Life at Death* (New York: Quill, 1980) 238–239.
130. Raymond A. Moody Jr.., , with Paul Perry. *The Light Beyond* (New York: Bantam Books, 1988), 14–15.
131. Monroe, *Journeys Out of the Body,* 265.
132. Talbot, 242.
133. Edgar D. Mitchell, DSc, , *Psychic Exploration: A Challenge for Science* (New York: G.P. Putnam's Sons, 1974), 368.
134. Talbot, 229 (Kenneth Ring, *Life at Death*)
135. McTaggart, 160.
136. McTaggart, 161.
137. Mitchell, *The Way of the Explorer,* 113.
138. Mitchell, *The Way of the Explorer,* 113.
139. Allan Snyder, DSc, , "Autism and Extraordinary Ability: Genius Locus," *The Economist,* April 16, 2009, http://www.economist.com/node/13489714.
140. "Autism to Brilliance: The Blake Cochran Story," *Caring Health Source,* Accessed January 10, 2012, http://www.caringhealthsource.com/autismtobrilliance.aspx.
141. Goswami, 88–89.
142. Lipton, *Biology of Belief.*
143. Lipton, *Biology of Belief.*
144. Myss, 2.
145. Myss, 2.
146. Gerber, 13.

147. Joe Dispenza, *Evolve Your Brain, The Science of Changing Your Mind* (Sydney: Nibbana, 2007), audio CD.
148. Braden, *The Science of Miracles.*
149. Talbot, 92.
150. Talbot, 93.
151. Peirce, 212.
152. Peirce, 213.
153. William Walker Atkinson, *Thought Vibration; or the Law of Attraction in the Thought World* (1906; reprint, n.p.: Kessinger Publication, 1996), 7–8.
154. Albert Einstein, *Science, Philosophy and Religion: A Symposium* (New York: The Conference on Science, Philosophy and Religion in Relation to the Democratic Way of Life, 1941), http://www.sacred-texts.com /aor/einstein/einsci.htm.
155. Mitchell, *The Way of the Explorer,* 163.
156. Jones, 87.
157. Peirce, 228.
158. Peirce, 118.
159. Peirce, 48.
160. Peirce, 106.
161. George Moore, *The Brook Kerith* (1916; reprint, n.p.: General Books LLC, 2009), 122.
162. Peirce, 47.
163. Martin Seligman, , *Authentic Happiness* (New York: Free Press, 2002), 113.
164. Seligman, 116.
165. Peirce, 100.
166. Peirce, 108.
167. Patricia Carrington, , *Discover the Power of Meridian Tapping* (Brookfield, CT: Try It Productions, LLC, 2009).
168. John Whitmore, *Coaching for Performance: GROWing People, Performance and Purpose* (London, UK: Nicholas Brealey Publishing, 2002)
169. Gerber, 104.
170. Gerber, 109.
171. Gerber, 109.
172. "24/7 Prayer in Washington State," *National Day of Prayer,* accessed January 11, 2012, http://nationaldayofprayer.org/news/answered-prayer/.

173. Emoto, 143.
174. Radin, 191–201.
175. Lynne McTaggart, *The Intention Experiment: Using Your Thoughts to Change Your Life and Your World* (New York: Free Press, 2007).
176. http://www.theintentionexperiment.com/.
177. Edgar Mitchell, DSc, "Psychic Exploration: A Challenge for Science," *Noetic Now* (June 2011), http://www.noetic.org/noetic/issue-eleven-june/psychic-exploration/.
178. "Goals & Objectives," *International Academy of Consciousness*, 2010, http://www.iacworld.org/english/academy/goals-objectives.

Glossary

Angel – A non-physical, interdimensional being, or energy, whose main purpose with regard to human beings is to guide and protect them, especially in relation to spiritual growth. Their exact character and appearance depend partly on the nature of their specific function and partly on the mindset of the person they are dealing with. Experiences associated with them range from seeing a vision or hearing a voice to sensing a presence or receiving a thought-message.

Akashic – *Akasha* is a Sanskrit term meaning "ether;" the Akashic records, which are stored in another dimension, are the non-physical record of all human soul events as well as the history of the cosmos. The idea was popularized through the teachings of the theosophists.

Atom – Once thought to be the smallest building block of matter, the atom's nucleus consists of one or more positively charged protons and neutral (having no electrical charge) neutrons, which are circled by negatively charged electrons. The number and type of these subatomic particles combine to form the different types of atoms, or elements, which in turn create various types of molecules (such as the familiar H2O molecule).

Aura – An energy field composed of various energetic layers (many models describe seven layers) surrounding the physical body. These fields are also known as "bodies" and are related to the chakras, which can be defined as energy centers or states of consciousness. According to many of the models, the physical body itself is considered to be a form of energy—though the coarsest and densest—that is penetrated and affected by the other fields.

Autism – Viewed as a disorder by many, autism usually becomes apparent before the age of three and is most characterized by variations of the following symptoms: linguistic development that is either slow or follows some unusual pattern; a lack of interest in peers and their activities; and unusual play patterns or self-stimulatory behavior, such as a child flapping or staring at his or her own hands. It has been defined as the inability to adequately filter the multitude of stimuli that surround us every minute. More recent research, however, has begun to connect autistic behavior to extraordinary abilities of various types, and it is thought that some autistics may perceive things at a higher frequency than most other people.

CERN – Originally standing for "Conseil européen pour la recherche nucléaire," CERN is short for the European Organization for Nuclear Research (Organisation européene pour la recherche nucléaire). The acronym stands for the largest particle physics laboratory in the world as well as for the organization that runs it. It is located on the border of France and Switzerland.

Chakra – A non-physical energy center or vortex sometimes described as resembling a flower, a wheel (Sanskrit *chakrum,* plural *chakra,* meaning "wheel"), or a stained-glass ball. Each chakra is associated with a different subtle body, frequency, or dimension. The seven major chakras most known in the West are aligned along a central column corresponding to the spine, beginning at the base and ending at the top of the head. They are usually designated (from bottom to top) as the Root, Sacral, Solar Plexus, Heart, Throat, Brow, and Crown. Each chakra is associated with specific characteristics as well as stages of consciousness and experience. The most commonly known of these are the colors of the rainbow, proceeding upwards from the Root (red) to the Crown (violet).

Connections – The underlying link between thought, energy, and frequency (or vibration) that produces the various phenomena that we experience, both explained and unexplained. Those who experiment with such things have noted how changes in patterns of thought or perception can produce palpable changes in experience, including what is called solid-state matter.

Consciousness – In its most basic sense, "consciousness" can be defined as awareness—either pure existence or perception. Different states of consciousness are thought to be tied to different subtle bodies and energy centers, and it is becoming more and more generally acknowledged that consciousness directly affects experience and is the source of existence itself.

Darwin Theory – Darwin's theory of evolution is based on the idea of a common ancestry for all life forms, specifically, that the complex evolves from the simple over a long period of time through a very slow process of natural selection based on those characteristics most suited for survival. The supposition is that this process can evolve not just a superior version of a species but an entirely new species, and that inferior species will eventually die out.

DNA – Deoxyribonucleic acid, a type of genetic code or blueprint found in virtually every cell of every organism. The information is stored in four chemical bases that form pairs and, together with a sugar and a phosphate molecule, create strands called nucleotides, which in turn form a double-helix (spiral) pattern. One of the important features of these DNA strands is their ability to replicate themselves. Another is that the blueprint furnished by our DNA is not the final word in our development: our cells are apparently able to turn the genetic information on and off, implying a degree of interaction with other factors.

$E=mc^2$ – Albert Einstein's famous equation, which translates as "energy equals mass times the speed of light squared." The equation is based on the idea that mass and energy are the same phenomenon in different forms.

Einsteinian – Relating to Albert Einstein (1879-1955), the 20th-century physicist whose work revolutionized physics. His most acclaimed achievements are his theory of relativity, expressed in his now famous equation $E=mc^2$, which changed our concepts of space-time; and his discovery of the photoelectric field, for which he won the Nobel Prize in Physics in 1921.

Electromagnetic Scale – Also known as the electromagnetic spectrum, it measures different levels of electromagnetic radiation according to frequency, wavelength, or photon energy. Levels range from the lowest known frequencies, such as those associated with touch or sound, to the highest, such as Gamma and Cosmic rays. In both physics and metaphysics, the term "octaves" is used to describe the different frequency levels, all the way up to the highest levels of Divine Consciousness.

Electron – A negatively charged subatomic particle that circles the nucleus of an atom. Its mass is approximately 1/1836 the mass of a proton.

Element – One of 118 chemically irreducible substances that, together with other chemical elements, constitute the material world. The chemical purity of each element results from its having only one type of atom, and the specific type of element—such as oxygen, hydrogen, carbon, iron, or gold—is determined by the number of protons in the atom's nuclei, which is also the element's atomic number.

Energetic Body – The non-physical subtle body or energy layer immediately outside the physical body and forming a part of a person's total aura. The vibrational patterns of this body are related to the structure and health of the physical body.

Energy – In both physics and metaphysics, energy is the latent, invisible (to the naked, unaided eye) aspect of mass. In physics, perhaps the most basic definition of energy describes it as the potential or actual ability of one physical system to exert force upon (to work upon) another physical system over a given distance. In metaphysics, it refers to the vibrational potency which precedes manifestation and which is highly affected by thought and emotion (intention).

Entanglement – The notion, which has been observed on both a microscopic and macroscopic scale, that particles created together are interconnected through space and time.

Extraphysical Consciousness – Consciousness that resides outside of and, potentially, apart from the physical body, as in out-of-body or near-death experiences.

Feng Shui – An ancient Chinese method of harmonizing people with their physical environment, not only through correct placement but through correct timing as well. It involves understanding and working with various ancient Chinese concepts, such as qi (energy); yin and yang, the polar opposite aspects of qi; the eight trigrams of the I Ching, or Book of Changes; the five Chinese elements and their cycles; and Chinese directional theory. Proper application of these and other laws of energy and placement can lead to increased health, happiness, and wealth.

Frequency – In physics, the number of oscillations of a wavelength within a given unit of time (as in cycles per second). (See also Electromagnetic Scale.) In metaphysics, the term "frequency" is often used to convey the quality of a person's energy as determined by their patterns of thinking and feeling.

GROW Model – A Life Coach problem-solving acronym which is thought to have been first developed by Sir John Whitmore, Graham Alexander, and Alan Fine in the 1980s. GROW stands for: Goal – Reality – Options – Way Forward. The Life Coaching Academy, founded in Australia, added the extra "I" and "N" (Issue and Nail Down) to the model to make IGROWN.

Immanent Energy – Pure energy, unaltered by human thoughts and emotions; also known as qi (pronounced "chi"), prana, vital force, etc., depending on the system.

Interconnectedness – The obvious extension of the notion of the underlying unity of all things, whether that unity is conceived of or experienced as consciousness or energy. The inseparability of the observer and the observed in quantum physics; the continuity of space and time; the connection between a single thought in one location and its effect in another location or time; the immutability of metaphysical laws, such as "as you give, so shall you receive"—all of these are examples of interconnectedness.

Intraphysical Consciousness – Consciousness that is experienced from within the physical body.

Invictus – In the original Latin, the word means "unconquered," "undefeatable." It has been used as a title for various things, including a poem, a movie, and two albums. In this book, it refers to the poem by William Ernest Henley, a Victorian poet who lived from 1849-1903. The poem exemplifies Henley's indomitable spirit in the face of adversity and has inspired numerous others, including Nelson Mandela.

Kirlian (also known as Kirilian) – a photographic process discovered by a 20th-century Russian electrician and his journalist wife, Semyon and Valentina Kirlian, that makes it possible to photograph the aura by applying an electrical charge to the object or by placing it in an intense electrical field.

Law of Attraction – The notion that we are what we think about, that whatever we hold strongly in thought and belief will ultimately manifest itself in our lives; also, that we have control over our thoughts and can therefore directly control our experience by changing our thoughts and attitudes.

Life Coaching – A relatively new, highly flexible field that focuses on helping a person achieve maximum satisfaction and growth in one or more areas of life. Rather than delving into the past, it achieves its ends by examining the person's present situation and then using a combination of tools and techniques—including focused, personalized goal-setting and the creation of a safe, positive, motivating environment—to help the individual experience an optimal life.

Matter – Generally considered to be the substance that constitutes the physical universe, matter is a concept that is still in flux as scientists continue to discover new dimensions of existence—from the subatomic to the cosmological. Models range from, among others, the familiar proton-neutron-electron structure to wave-particle theories to such little understood phenomena as dark matter and dark energy that defy the usual laws. Although matter is generally associated with mass, the more recent approach in physics that recognizes mass and energy as being equivalent gives us a different perspective, one that has long been recognized in metaphysics. The more advanced (and, in some cases, secretive) metaphysical systems speak of matter from different dimensions and matter as an illusion, the product of human concepts.

Metaphysics (Science Beyond the Physical) – Literally, "beyond physics," metaphysics has also been termed "Divine Science," and it can be described as the systematic study of spiritual and mental phenomena beyond the realm of the physical senses. At its highest levels, it is the study of essential and absolute Reality. In general, it is concerned with cause and effect at much subtler levels of existence than are commonly explored in the physical realm, though this has begun to change in the last hundred years, especially in the realm of physics. Metaphysics is closely linked with spirituality.

Molecule – Two or more atoms joined together. The atoms can be of one type only, as in the case of H_2, or they can be of two or more types, as in H_20. Molecules with more than one type of atom are called compounds.
Morphic Fields – What Rupert Sheldrake calls the "invisible organizing fields" that underlie the development of individual "morphic units." Their connection to the larger "morphic field," similar to Jung's "collective

unconscious," enables these smaller "morphic fields" to transmit information to the whole and, thus, to other members of the same or different species without the need for physical contact or specific knowledge of each other.

Morphic Resonance – Another Sheldrake term, which he defines as "non-local resonance" developed by repeated patterns of behavior and transmitted through the "morphic field." Sheldrake maintains that what we have deemed "natural laws" are not actually laws but habits transmitted and reinforced to the species through this resonance-field connection. In more specific instances, "morphic resonance" is what enables a particular "morphic unit" with a strong resonance of a given type to tune in non-locally and without technological aid to a similarly oriented individual or collective, and vice-versa.

NDE (Near-Death Experience) – An out-of-body experience of consciousness, in which the body actually exhibits signs of death while the soul, or consciousness, experiences some of the transitional phases between this and alternate dimensions. In NDEs, the soul is sent back and/or chooses to return, and at that point, the seemingly dead body revives.

Nucleus – The dense center of the atom, containing the protons and neutrons.

NLP (Neuro-Linguistic Programming) – Founded by Richard Bandler and John Grinder in the 1970s, NLP seeks to help a person improve the connection between thinking, language, and behavior for the purpose of achieving optimal results in life. Its premise is that by modeling those who habitually produce excellence, others can do so as well.

Newtonian – Pertaining to Sir Isaac Newton (1643-1727), whose discoveries in the fields of mathematics, optics, and physics dominated scientific thought for three hundred years. A polymath who studied and wrote on theology and philosophy as well as science, he is best known for his significant advances in mathematics, optics, and physics, including his laws of motion and his theory of gravity.

Noetic Science – The application of scientific methods to the study of what has been termed "subjective" experience (metaphysics, spirituality, intuitive knowing, etc.). It is distinct from the branch of philosophy called "noetics," which is the study of the mind or intellect.

Non-locality – Refers to consciousness that is unbounded by the limitations of physicality.

Observer – Refers to the "observer effect," which states that the observer and the thing observed are not separate but interconnected phenomena. The very presence of the observer, either in the form of a measuring instrument or a consciousness, acts upon the thing observed and, in the process, changes it. This concept has been applied in relevant ways to various spheres of physics as well as the social sciences and metaphysics.

OBE or OOBE (Out-of-Body Experience) – Similar to a Near-Death Experience in that the consciousness, or soul, separates from the body, at times witnessing the body from a distance outside or above it. The difference, of course, is that the body does not show signs of physical death.

Paranormal – Literally, beyond or outside the normal scope of things. Paranormal subjects are those not readily accepted by science because they defy its current capacity to measure them or prove their existence. Exactly what the term includes is changing as new instruments and techniques are developed and new ways of thinking and seeing explored.

Particle – In general terms, a minute amount, object, or fragment. In physics, the terms refers to subatomic particles, such as electrons, quarks, and photons (examples of elementary particles) or baryons (composite particles), such as protons and neutrons.

Planck – Max Planck (1858-1947), winner of the Nobel Prize in Physics in 1918 for his discovery of quantum physics. His name has since been attached to various aspects of quantum physics, such as the Planck unit, Planck scale, Planck length, Planck mass, etc., which describe this minute subatomic world.

Quantum Mechanics/Physics – The area of physics dealing with matter at the atomic and subatomic levels. The susceptibility of subatomic particles to the "observer effect" and their tendency to act in ways that defy the logic of macroscopic theories and observations has challenged our commonly held notions of matter and reality.

Quantum Theory – Another term for quantum physics or mechanics; the theory. Originated by Max Planck in the early 1900s, it includes the ideas that matter consists of vibrating energy packets called "quanta" and that subatomic particles are capable of functioning as both a wave and a particle as a result of the "observer effect."

Reality – In its simplest definition, reality is the term used to denote being, or existence. Exactly what that means to us varies with our perception and understanding. If you come from the perspective of "you have to see it to believe it," you will experience your world or existence differently from those who accept the notion that "what you believe is what you see." Aside from the fact that both science and history have disproved the first idea again and again (think of Columbus and the "earth-is-flat" mentality), those who have experimented with the second (again, Columbus is a great example) have made breakthroughs that have changed the way the average person views the world.

Resonance/Resonation – The matching of the natural frequency of an object or entity to equivalent or harmonious frequencies in its environment and according to whatever parameters are available (in the case of sound, membranes, cavities, etc.). Metaphysically, "resonance" or "resonation" refers to the idea that a person or other being can, either locally or non-locally, tune in and link to those things, situations, and people that echo the individual's or group's own state of being.

RNG/REG – Random Number Generator/Random Event Generator. Random Number Generators are machines which have been designed to generate seemingly random numbers and events using the laws of quantum mechanics as their basis. Random Event Generators are the same hardware device used to measure the effects of consciousness on random events.

Spirituality – Closely associated with metaphysics because it pertains to dimensions beyond what is termed the material world, spirituality concerns itself with absolute Truth, or essential being, often called God; with states of consciousness far beyond the "norm;" with the return to the true state or authentic Self; and with the ongoing improvement and clarification of character, thought, and purpose, ultimately resulting in an expanded sense of being and self. Spirituality is further often related to healing, altered or expanded perception, and the awakening of previously inaccessible faculties and abilities.

Superposition – The ability of a particle to exist in more than one location or state at a time. The implications of superposition on a macroscopic scale have long been known, in the form of bilocation and multilocation, to advanced spiritual practitioners from both Eastern and Western traditions.

Synchronicity – The simultaneous or seemingly coincidental occurrence of related events, or events considered fortuitous, in ways not easily explainable according to the usual macroscopic views and theories of reality.

Telepathy – The ability to transmit and receive messages in the form of thoughts, feelings, visions, and physical sensations through extrasensory means, often to or from a non-local source.

The Field – A unified field of vibrating energy or consciousness that is the basis of connectedness between all phenomena. Long confirmed by advanced spiritual practitioners, this idea is still being worked on by physicists and other scientists who historically have used different models to explain and describe different types of phenomena, ranging from the cosmological to the subatomic. The notion of existence as a unified, interactive whole rather than as a multitude of separate parts, together with the acknowledgment by some scientists of phenomena that defy explanation according to previous models, is leading to new approaches, perceptions, and conclusions in a number of different fields of endeavor.

The Talmud – One of the central texts of Judaism. Its two main components are the Mishnah, interpretations of Hebrew scripture and law compiled around 200 A.D.; and the Gemara, later rabbinic commentaries on the Mishnah. It also contains non-legal teachings known as Aggadah that include Jewish folklore, allegory, history, morality, and theology.

Thought – Mental activity, including logical deduction, imagination, hunches, images, etc. Thoughts that are repeated—especially if they are organized and directed in some way—often take on the power of conviction or belief and tend to influence our experience, regardless of whether they have any rational basis or not. Both metaphysical experience and scientific observation of anomalies have shown that thought has the ability to produce immediate, palpable physical, psychological, and experiential changes.

Tuning Fork – A two-pronged acoustic resonator, usually made of steel, that produces a clear fundamental frequency (minus excessive overtones) by striking it against an object; a metaphor for consciousness or state of mind.

TFF (Tuning Fork Frequency) – The specific frequency of a tuning fork depending on its individual parameters, including the length of the prongs and the material from which it is made. The idea of the TFF as a metaphor for individual or collective consciousness is based on the notion that our thoughts, emotions, and actions resonate at a certain frequency, and that the repetition of any of these, especially if it proceeds from deep conviction, naturally and inevitably resonates with and potentially activates similar frequencies.

Vacuum – Originally thought to be a state empty of matter; a void. According to both physicists and advanced mystics and spiritual practitioners, however, there is, in fact, no such thing. In physics, the so-called vacuum, void, zero-point field, or ground state has been found to contain energy in the form of fleeting subatomic particle and wave activity.

Vortex – A spiraling, whirling mass of matter (water, wind etc.), as in a tornado or a whirlpool, that can act as a strong sucking or attracting force; a metaphor for a similar non-physical experience; a non-physical phenomenon of a similar type in alternate dimensions (chakras, for example).

Wave/Particle Duality – The recognition that a quantum particle can act as either a wave or a particle. This was determined in the dual-slit experiment conducted by quantum physicists, in which subatomic particles left wave patterns until observed in motion, when they acted like particles.

Bibliography

"About Anna." *Anna Meares. 2010.* www.annameares.com.au/about-anna.html.

Andrews, Ted. *How To See and Read the Aura.* Saint Paul, MN: Llewellyn Publications, 1991.

Atkinson, William Walker. *Thought Vibration; or the Law of Attraction in the Thought World.* 1906. Reprint. N.p.: Kessinger Publications, 1996.

"Autism to Brilliance: The Blake Cochran Story." *Caring Health Source.* Accessed January 10, 2012. http://www.caringhealthsource.com/autismtobrilliance.aspx.

Bohm, David. *Wholeness and the Implicate Order.* London: Routledge & Kegan Paul, 1980.

Braden, Gregg. *The Science of Miracles: The Quantum Language of Healing, Peace, Feeling, and Belief.* New York: Hay House, 2007.

Brennan, Barbara Ann. *Hands of Light: A Guide to Healing Through the Human Energy Field.* New York: Bantam Books, 1988.

Byrne, Rhonda. The Secret. Web-based feature film. Directed by Drew Heriot. 2006. N.p.: Prime Time Productions, 2006. DVD.

Carrington, Patricia. *Discover the Power of Meridian Tapping: A Revolutionary Method for Stress-Free Living.* Brookfield, CT: Try It Productions, LLC, 2009.

Dispenza, Joe. *Evolve Your Brain, The Science of Changing Your Mind.* Sydney: Nibbana, 2007. Audio CD.

Emoto, Masaru. *The Hidden Messages in Water.* Hillsboro, OR: Beyond Words Publishing, 2001.

Gerber, Richard. *Vibrational Medicine for the 21st Century: A Complete Guide to Energy Healing and Spiritual Transformation.* London: Judy Piatkus (Publishers), 2000 (republished as *A Practical Guide to Vibrational Medicine: Energy Healing and Spiritual Transformation.* New York: Harper Collins, 2000).

Gilder, Louisa. *The Age of Entanglement: When Quantum Physics Was Reborn.* New York: Knopf, 2008.

Goswami, Amit. *Physics of the Soul: The Quantum Book of Living, Dying, Reincarnation, and Immortality.* Charlottesville, VA: Hampton Roads Publishing Company, 2001.

Hawking, Stephen and Leonard Mlodinow. *The Grand Design: New Answers to the Ultimate Questions of Life.* London: Bantam Press, 2010.

Hay, Louise. *You Can Heal Yourself.* Carlsbad, CA: Hay House, 1999.

Hicks, Esther and Jerry. *Ask and It Is Given.* Carlsbad, CA: Hay House, 2004.

———. *The Vortex: Where the Law of Attraction Assembles All Cooperative Relationships.* Carlsbad, CA: Hay House, 2009.

Jones, Marie D. *Modern Science and the Paranormal (Haunted: Ghosts and the Paranormal).* New York: Rosen Publishing, 2009.

Kaku, Michio. *Hyperspace: A Scientific Odyssey Through Parallel Universes, Time Warps, and the 10th Dimension.* New York. Oxford University Press, 1994.

Lipton, Bruce H. *Biology of Belief: Unleashing the Power of Consciousness, Matter & Miracles.* Boulder: Sounds True, 2006. Audio CD.

Matheson, Richard. *What Dreams May Come.* Feature Film. Directed by Vincent Ward. N.p: Interscope Communications, 1998.

McTaggart, Lynne. *The Field: The Quest for the Secret Force of the Universe.* London: Element, 2003.

———. *The Intention Experiment: Using Your Thoughts to Change Your Life and Your World*. New York: Free Press, 2007.

Mitchell, Byron Kathleen and Stephen. *Loving What Is: Four Questions That Can Change Your Life*. Ojai, CA: Harmony Books, 2002.

Mitchell, Edgar D. *Psychic Exploration: A Challenge for Science*. New York: G.P. Putnam's Sons, 1974.

———. *The Way of the Explorer*. Franklin Lakes, NJ: New Page Books, 2008.

Monroe, Robert A. YouTube. 1979 Interview Courtesy WPIX San Francisco. http://www.youtube.com/watch?v=wrLApcABHQw&feature =related, October 27, 2009.

———. *Journeys Out of the Body*. New York: Doubleday, 1977.

———. *Ultimate Journey*. New York: Doubleday, 1994.

Moody, Raymond A. with Paul Perry. *The Light Beyond*. New York: Bantam Books, 1988.

Myss, Caroline. *Energy Anatomy*. Boulder, CO: Sounds True, 1997. Audio Cassettes and Booklet.

Peirce, Penney. *Frequency: The Power of Personal Vibration*. New York: Atria Books, 2009.

Radin, Dean. *Entangled Minds: Extrasensory Experiences in a Quantum Reality*. New York: Paraview Pocket Books, 2006.

Ring, Kenneth. *Life at Death*. New York: Quill, 1980.

Rothman, Tony and George Sudershan. *Doubt and Certainty: The Celebrated Academy Debates On Science, Mysticism, Reality*. Cambridge, Massachusetts: Helix Books, 1998.

Segal, Inna. *The Secret Language of Your Body: The Essential Guide to Healing.* Glen Waverley, VIC: Blue Angel Publishing, 2007.

Seligman, Martin. *Authentic Happiness*. New York: Free Press, 2002.

Sheldrake, Rupert. *Dogs That Know When Their Owners Are Coming Home: And Other Unexplained Powers of Animals*. London: Hutchinson, 1999.

———. *The Presence of the Past: Morphic Resonance and the Habits of Nature*. London: Collins, 1988.

Snyder, Allan. "Autism and Extraordinary Ability: Genius Locus." The Economist, April 16, 2009. http://www.economist.com/node/13489714.

Spirit Space: A Journey Into Your Consciousness. Directed by Braden Barty, 2008. N. Hollywood, CA: Wirewerks Films, 2008. DVD.

Talbot, Michael. *The Holographic Universe*. New York, NY: Harper Perennial, 1991.

Taylor, Jill Bolte. "Jill Bolte Taylor's Stroke of Insight." Video. TED: *Ideas Worth Spreading*. 2008. http://www.ted.com/talks/jill_bolte_taylor_s_powerful_stroke_of_insight.html.

Taylor, Jill Bolte. *My Stroke of Insight: A Brain Scientist's Personal Journey*. London: Hodder & Stoughton, 2008.

"The Electromagnetic Spectrum" (Illustration). *National Aeronautics and Space Administration*. Last updated January 6, 2011. http://mynasadata.larc.nasa.gov/ElectroMag.html.

The Quantum Activist (with Amit Goswami, Ph.D.). Directed by Ri Stewart and Renee Slade, 2009. Eugene, OR: Blue Dot Productions, 2009. DVD.

The Voice. Directed by David Sereda, 2008. N.p.: Voice Entertainment, 2008. DVD.

Tolle, Eckhart. The Power of Now: A Guide to Spiritual Enlightenment. Sydney, NSW: Hodder Australia, 2004.

Tolle, Eckhart. The Power of Now: A Guide to Spiritual Enlightenment. Novato, CA: New World Library, 1999.

Vieira, Waldo. Projectiology: A Panorama of Experiences of the Consciousness Outside the Human Body. Rio de Janeiro, Brazil: International Institute of Projectiology and Conscientiology, 2002.

What the Bleep!? Down the Rabbit Hole (Originally What the Bleep Do We Know!?). Directed by William Arntz, Betsy Chasse, and Mark Vincent. 2004. N.p.: Roadside Attractions, 2006. DVD.

Whitmore, John. Coaching for Performance: GROWing People, Performance and Purpose. London, UK: Nicholas Brealey Publishing, 2002.

Index

F

G

H

Q

R

S

T

U

W

About the Author

Hayley Weatherburn is fascinated by energy, more specifically, the frequencies in our world pertaining to human energy and how they affect us as individual consciousnesses.

Having always been interested in the power of the mind, Hayley is a lifelong learner. Questions surrounding the science behind thought, energy, and frequency plague her, so she continues to read, research, and educate herself on all aspects of consciousness, unexplained phenomena, quantum physics, and anything else which helps her to understand this fascinating reality we experience on a day-to-day basis. In other words, anything that can help explain the unexplainable!

Not only has Hayley written about these phenomena, she also enjoys giving talks and seminars on metaphysics which are aimed at helping people reach their true resonating frequencies. Her goal in life is to inspire people to evolve on a conscious level by helping them discover their unique traits.

Hayley spent five years of her life travelling through Europe, America, Asia, and Africa, learning from all of her experiences on the way. Finally, however, she decided to settle in Brisbane, Australia. When she isn't writing and researching, she is travelling, practicing Bikram Yoga, volunteering for the International Academy of Consciousness, or generally just socializing with her close friends and family.